SECURITY GUARD

A Guidebook
For
* Guards
* Officers
* Managers
Of
Agency And In House
Security Forces

DAVID Y. COVERSTON
Illustrated by Anne C. Brown

SECURITY SEMINARS PRESS

Ocala, Florida

Library of Congress Cataloging-in-Publication Data
Coverston, David Y. (David Yost), 1920–
    Security guard.

    Includes index.
    1. Police, Private — United States. 2. Industry — United States —
Security measures. I. Title.
HV8290.c68 1986   363.2'89'0973   85-30430
ISBN 0-936101-01-6
ISBN 0-936101-00-8 (pbk.)

Published by Security Seminars Press (Publishers)
1204 S.E. 28th Avenue
Ocala, Florida 32671

Published in the United States of America

CONTENTS

CHAPTER I

For Openers
1

CHAPTER II

The Job
17

CHAPTER III

The Professional
31

CHAPTER IV

Legal Guidelines
45

CHAPTER V

Criminal Acts
57

CHAPTER VI

Hazardous Humans
65

CHAPTER VII

The Worst Criminals
83

CHAPTER VIII

Guns
93

CHAPTER IX

Natural Hazards
101

CHAPTER X

Outside Areas
111

CHAPTER XI

Alarms
121

CHAPTER XII

Physical Controls
131

CHAPTER XIII

Special Situations
141

CHAPTER XIV

First Aid
159

CHAPTER XV

Fire
169

CHAPTER XVI

Put It In Writing
177

ACKNOWLEDGEMENTS

The author gives his heart felt thanks to the many people who gave so much of their time and help in getting this book into print.

Extra special gratitude is expressed to the following for their valuable advice and other contributions.

To my daughter, and illustrator, Anne C. Brown, for taking time from her drafting and map making work and the rearing of my grandchildren, to make many of these pages come alive.

To my daughter, Lucinda, for keeping track of and assembling pages, illustrations and other materials for the final draft.

To my brother, Sam, for his suggestions, critique of copy and assistance in the advertising campaign.

To William L. "Bill" Meadows for giving me the idea for the book and for his professional input.

To Betty Wright, President of the National Association of Independent Publishers, for her words of wisdom and guidance.

To Nancy Mahrer for her typing and word processing as well as her knack for making the manuscript more readable.

And, finally, to my wife, Martha, for her patience, understanding, and constant encouragement.

DEDICATION

During a war the enemy is bombed from the air, shelled from the sea, and blasted with guided missles. Then, with fixed bayonet, comes the G.I., taking possession. The aircraft, the warship and the rocket have taken their toll — but it's the G.I. on the final firing line.

In so-called peaceful civilian life, building halls and lobbies are monitored by closed circuit television, computer centers are electronically wired, smoke detectors abound, and movement sensors are properly located. The TV screen discovers an unauthorized entry; an alarm sounds from the computer area; the smoke detector indicates a fire in progress, and the movement sensor locates the unauthorized entry in the computer area. The modern security system has completed its job — but it's the Security Person with drawn weapon, if armed, or with night stick, on the final firing line.

The two action scenes are a great deal alike, the final scenes are a great deal different. Let's compare.

The G.I.'s have been well *trained*. They move in, knowing what to expect, secure the objective, and turn it over to occupation forces. They retire from the scene, and after receiving their well-earned medals and citations, return home to a grateful citizenry.

Security Personnel have usually received little or no *training*. After catching the criminal intruder and holding for the police, help is given in cleaning up the premises before standing by for relief. After being relieved of duties, the required report is written, a note made that testimony will be required at a later date, then to home. Thanks may be received, or may not be received, and the tragic fact is, other than for those at the scene, few people will ever know the part Security Personnel played in the affair.

It is to these almost unknown and seldom praised men and women this book is dedicated.

PREFACE

TWO V.I.P.'S

By conservative estimates, the Security Profession employs more than a million men and women in the U.S. The number of Security personnel employed is more than double the number of all law enforcement officers employed by all of the police departments and sheriffs' offices in the fifty states. These numbers make valid the designation of the Security Profession as a VERY IMPORTANT PROFESSION and the designation of Security Personnel as VERY IMPORTANT PERSONNEL.

Projections made on a nationwide basis to the year 2000 indicate there will be greater need in the future for *trained* Security Personnel. Regardless of the economy, as the population grows, so do the number of muggings, rapes, thefts and other forms of crime the criminal element inflicts upon the public.

In spite of the hard work and dedication of the Sheriff, the Police, State Troopers and the F.B.I., the criminal is gaining ground. Law enforcement agencies need help. Their officers are public employees and cannot be expected to work solely for private interests. As a result, private interests are going to have to provide much of their own security in the days ahead.

The major source of help needed by public law enforcement agencies, and for the help needed by the private sector in the fight against crime, is the Security Profession with *trained* forces of Security Personnel. The Security Profession is growing rapidly and the demand for *trained* Security Personnel is never satisfied.

The citizen, business places, industries and other establishments face many and varied hazards. To combat these hazards requires the installation of more and better security measures. In addition, the complexity of many hazards requires better *trained* Security Personnel.

Arson, vandalism, shoplifting, burglary, assault and other crimes committed against people and property are widespread and growing. To slow this giant cancer in our society, we must move to prevent criminal acts. The first line of prevention is provided by the Security Profession and Security Personnel.

Many community crime prevention programs rely upon the use of Security Personnel on beats, motor patrols and standing posts. A number of retirement communities, condominium associations and apartment complexes employ Security Personnel to help prevent crimes in their areas.

The inside criminal, using positions of trust, embezzles and otherwise steals property on an ever increasing basis. And, because an enterprise must not only earn assets but must keep these assets, it must devise methods to safeguard them. A properly installed security system with adequate numbers of *trained* Security Personnel is the best prevention. The installation of such security costs money, but a great portion of the cost is offset by reduced insurance premiums. Additionally, keeping earned assets previously stolen makes security expenditures self supporting.

Cities, Counties and States are drowning in crime and criminal activities. Evidence the crowded jails and prisons in our cities, counties and states. As a result, the individual, business places, industry, and corporations must bear greater responsibility for the protection of personal and property assets. They must turn to the only source of help — The Security Profession and Security Personnel.

When considering the fact crimes against business exceeds fifty billion dollars a year — that two thirds of small business concerns fail following a fire — and that crimes against persons are the highest in the history of our country — it is easy to see how much we need the Security Profession — a VERY IMPORTANT PROFESSION — and Security Personnel — VERY IMPORTANT PERSONNEL.

SECURITY OFFICER CODE OF ETHICS

As A Security Officer, I Pledge . . .

To Discharge The Obligations And Responsibilities Of My Position In The Best Interest Of My Employer, Except Where Such Would Conflict With Law Or Morality.

To Perform My Duties To The Full Extent Of My Ability, Placing The Protection Of My Employers' Property Above My Own Interest.

To Safeguard The Lives Of Persons Under My Protection, Even At The Risk Of Personal Danger.

To Observe The Highest Principles Of Honesty, Integrity, And Loyalty At All Times.

To Respect, Obey, And Uphold The Law, And To Assist Peace Officers In The Performance Of Their Duties, If Requested To Do So Or In The Event Of An Emergency.

To Remain Constantly Alert, Observant And To Be Physically And Mentally Prepared.

To Report Accurately, Truthfully, And Promptly On All Matters Within The Scope Of My Authority, Without Regard For Friendships, Prejudices, Or Personal Advantage.

To Deal With The Public In A Businesslike Manner, Combining Firmness, Courtesy, Fairness, And Tact.

To Treat My Fellow Security Officers With Respect, Consideration, And Loyalty, Never Allowing Personal Feelings To Endanger The Work Relationship.

To Maintain A Neat, Well Groomed Appearance At All Times, And To Keep My Uniforms And Equipment In Proper Order.

To Take Advantage Of All Training And Educational Programs That Might Enable Me To Improve My Professional Ability.

CHAPTER I

FOR OPENERS

A LITTLE HISTORY

Literature is filled with references to the Security Profession and Security Personnel. History books, bibles, fictional works and professional journals are sources of such references. The daily and weekly newspaper, television shows and motion pictures have items that include the Profession and its Personnel.

Almost every newspaper edition has a story in which security is the topic and countless numbers of television viewers hear the actor or actress in danger utter the words, "Call Security".

The Bible has several references to Security including these:

Psalm 127 —   "The Watchman Waketh, But In Vain."
Isaiah —   "Go, Set A Watchman, Let Him Declare What He Seeth."

Ezekial —   "But If The Watchman See The Sword Come, And Blow Not The Trumpet, And The People Be Not Warned; If The Sword Come And Take Any Person From Among Them, He Is Taken Away In His Iniquity; But His Blood Will I Require At The Watchman's Hand."

Shakespeare, in four of his plays, wrote of the watch:

Richard III —   "Good Norfolk - Use Careful Watch, Choose Trusty Sentinels."

Othello — "Iago, We Must To The Watch."

Henry IV —   "The Sherife And All The Watch Are At The Door."

Anthony and
Cleopatra —   "Let's See If Other Watchmen Do Hear What We Do."

History books tell of guard forces in ancient times — notably the PRAETORIAN guard force of Rome, and select personnel from the military legions of Egypt known as the MAMELUKES.

Many present day officials in law enforcement have their origins in practices centuries old. A prime example is the law enforcement leader now known as the *Sheriff*.

In the earliest Anglo-Saxon system of protection, dating back many centuries, every ten families in an area were forcibly organized into a community law enforcement group bearing the name *Tithing*. The *tithing* responsible for safety of the area chose a leader. He became known as a *tithingman*. He represented the ten families. When the number of *tithingmen* reached ten, one of them was chosen to speak for all of the *tithings.* He was known as a king's *Reeve.* He represented one hundred families.

After an area had accumulated four or more hundreds of families, the area was designated a *Shire.* This was for protection and governmental purposes, and a single *Reeve* from the *Shire* was chosen as spokesman to the King. He could order the support of all in the *Shire,* for purposes of defense, or in searching for a criminal. In this capacity he became known as the *Shire Reeve.* Over the years, changes in language and coinage of words has resulted in the present day word of *Sheriff.*

Another modern official, all important in England, to less extent in the U.S., has a name taken from ancient English protective services. The *tithingman* was replaced, after some 700, years by an officer known as the *Constable.* The *Constable* was a high ranking officer in the household of the king. His name was a combination of the Latin words *"Comes Stabuli"* which roughly translated means *"Count of the Stable."* Over a period of time, the *Constable* in the palace became very important and powerful. He was usually given command of the king's armies when the king was away. Each hundred families had a *Constable* responsible for protecting the public from crime and criminals.

Outside of the king's household, appointments were made in the city much like the *Constable.* These men had the duty to oversee watchmen and keep peace. This appointee was known as *Captain of the Watch.* (We know him today as the *Chief of Police.*)

During the time of the *Constable,* a group of protectors came into being as city watchmen and (later) street musicians. Known as *Waytes* or *Waits,* their title came from Old English. *"To Make Wait"* meant "to keep watch," *"To Have In Wait"* meant "to be carefully on the watch." (Modern use is, a criminal "lies in wait.")

The *Waits* used horns and trumpets to sound off at

intervals to assure the population they were on duty and "all was well." Their becoming street musicians was probably their efforts to relieve themselves of boredom by playing their instruments and their need to pick up a little money through concerts.

During the reign of King Edward I, the Statute of Winchester (which was adding up all of the effective law enforcing experiences to that time), established a regular system of night patrols in towns. Boys from ages sixteen to twenty-one were required to take turns in patrolling the community. A *Constable* was appointed to supervise the boys. Although the boys were unpaid, the system worked in smaller towns. Public opinion was on the side of the boys and offenders were promptly dealt with. The victims were given sympathy and help; the criminal was removed from society for having taken away the rights of the victim.

During this period of time, one method of obtaining help from the citizenry was a process called *Hue and Cry*. This was a general alarm raised whenever it was found that a felony was being committed. When the alarm sounded, the *Constable* ordered everyone to go after the felon or to be fined and/or punished. Although it was a simple method of arrest, it was effective at the time. It helped in the finding and capture of the criminal.

During the 1500's in England, the men on street patrol were known as *"The Watch."* They were highly successful. But with the growth of their communities and the growth of crime, the job of being an unpaid policeman was no longer a source of pride. The citizen who could afford it hired someone to take his turn at patrol or watch. The low pay resulted in a loss of quality and forced King Charles II into hiring night watchmen.

The watchmen hired, armed only with staff and a bell, were known as *Charlies*. They were *untrained*, poorly paid, and the target of every criminal at large in

6

London. They struggled against the odds for a long period of time. But, the lack of support, lack of *training*, lack of control and poor supervision turned them from preventing crime to participating in crime. Shortly after the beginning of the 19th century the English people finally had enough and installed the system they have today.

During the one hundred fifty years Americans were ruled by the British they learned a great deal about taking care of themselves. They approved of the idea of protection. They believed keeping law and order was an individual's duty, that each person was responsible for others in society.

In Colonial America the forms of protection were little changed from those in England. The Constable was the responsible lawman in the towns. The Sheriff was responsible for the county.

There are few records of the Security Profession and Security Personnel during this period of time. However, there are a few highlights from that era.

— New Amsterdam, now New York City, had a citizen's watch for a time.

— Boston formed a nightwatch in its early history. It existed in some form or other for two hundred years before police were hired for day time duties.

The first fifty years following the founding of St. Louis, the only Security available was the night watch force.

In early America, watchmen were hired only when the populace felt threatened by an Indian attack or rioting by slaves. In those instances, a night of work, starting around 9 p.m., and lasting until light of day, was offered for a fifty cent fee for the night. When the threat of attack or riot was gone, even that small fee was not paid, and citizens were drafted for watch duty.

The drafted watchman status was one to be avoided if at all possible. The watchman had to face hostile Indians, hoodlums, fire, wolves, bears, wild dogs, runaway slaves, escaped convicts and other undesirables.

In addition the watchman had to put down disturbances, arrest drunks, break up fights, and question those on the streets after curfew. In some cases the watchman had other duties. They included filling, lighting, and extinguishing town lights, calling out the time at set intervals, and announcing weather conditions.

These drafted watchmen worked only after dark. They worked during the day at other jobs to support themselves and their families. There were no daytime police for protection of citizens or property. Everyone who could do so hired someone to serve for them. Eventually, the only people serving as watchmen were the poor who could be drafted and paid in tax rebates rather than cash.

As America grew from a nation of hunters and farmers into a nation of industrial and merchantile centers with seaports and transportation hubs, the protection problem grew. The concept of taking care of one another as a citizenship responsibility died. The watchmen of the time made brave attempts to cope with the growing crime problem, but the odds were against them. Their failures were due to lack of *training* and legal support rather than personal shortcomings.

Following the Revolutionary War and the birth of the U.S., the country grew so fast and the breath of freedom was so fresh there was no real effort to form any type of regular policing. Riots, terroristic mobs and gang fights were common to the cities. However, the idea of organizing and *training* a force of lawmen with authority and legal stature seldom came up.

Many cities put in a daytime force of so-called police

officers while retaining their night Watchmen. It was expensive; it was not efficient, and it caused friction within the communities. It was years before these cities began to correct the situation.

With the outbreak of the War Between the States there was a merging of the Watchman and the Police of that day. The Guard became a Policeman, as we define them today and the Police became what we define as Detectives today.

Following the civil conflict, the nation expanded. With it came outlaw gangs, Indian agitators, rustlers and others with the intent to take the law into their hands. Indians were used as lawmen for the next quarter of a century, working under Indian agents, being used mainly to guard railroads, mines, cattle ranges and logging operations. The Indian faded from view by 1900. Uniformed guards were used at the Exposition in St. Louis and the Exposition in Portland, Oregon. The normal procedure was to make watchmen out of employees who had little or no qualifications for the job.

During the years immediately before the U.S. entered World War I, many fires and explosions took place that were blamed on enemy agents. The 1916 explosion that destroyed a munitions plant located in Jersey City and the 1917 explosion of a munitions ship docked at Halifax, Nova Scotia were thought to be acts of saboteurs. These events, plus others, frightened businesssmen into hiring security forces for the duration of the war. After the war, interest in *trained* security waned. Industrial plants returned to the practice of filling security departments with retired messengers, janitors, and other employees no longer useful in other jobs.

The twenty years between World War I and World War II was a period with little activity in the Security Profession. There were crimes and criminals, many spawned by prohibition. However, the nation enjoyed prosperity the first ten years of this time and no one worried about security. The latter ten years of this time, the nation suffered an economic depression, and premises security of idle plants was turned over to custodial crews. Security Personnel were hired and used by management in many cases, to protect industrial plants which were being damaged by striking union members.

The Security Profession, as we know it today, and the Security Personnel who make it go, owe their existence to the Japanese. It was that day of infamy, 7 December 1941, when bombs fell on Pearl Harbor, that modern security was born.

Millions of men and women joined or were drafted into the armed forces. The concentration of industry and manufacturing was in large cities. Many new military camps, naval bases, and air fields were built around the nation. The need for security measures to protect life and property became all important. Thousands of men and women were hired into the Security Profession and the Watchman finally passed into oblivion.

After World War II, cities grew larger, agriculture had fewer and larger farms, and legislatures enacted more complex laws. Life became more complicated and the criminal element grew in numbers and sophistication. The problems of law enforcement officials ballooned out of proportion and Security Personnel became Very Important Persons.

Today we need the Security Profession and Security Personnel more than we ever have in our history. Criminals seem to have more rights than victims, lawyers seem more intent on helping the criminal than seeking justice for the law abiding. The light sentences judges hand criminals combined with the leniency of parole commissions have all but blotted out the criminals' fear of punishment.

THE PUBLIC IMAGE

The image the public has of Security Personnel is so-so. Over a period of years it has suffered, due to the "don't care" attitude, the poor personal appearance of the Personnel and their lack of training. None of these are excusable, particularly the "don't care" attitude.

However, there are many places where Security Personnel are highly respected and are considered important members of the parent organization. When we examine the reason, we find Security Personnel who have the proper overall attitude.

Security Personnel are responsible for the image of the profession. And, the treatment the public gives Security Personnel depends upon that image.

A negative, or bad impression left with the public is caused by slovenly actions, surliness, rudeness, a lazy and slouchy appearance, dirty or unpressed uniforms, knowing nothing, and not caring. In three words — a poor attitude.

A positive, or good impression left with the public is brought about by being neat, sharp in appearance, courteous and polite, businesslike and alert, knowledgeable and helpful. In three words — a good attitude.

Anyone who wants to become a good Security Person and climb the ladder of success must aid in getting rid of any poor image left by others. Regardless of education and background, Security Personnel should be proud of themselves. They should study their profession, keep themselves ready for inspection at all times and leave the impression they *are somebody.*

THE TRAINING SITUATION

It is a sad commentary on the profession, but in far too many instances *training* of both new and seasoned Security Personnel is neglected. It is not unusual for

applicants for Security Personnel jobs to be sent to be fingerprinted, and upon return, pending the issuance of a license, to be put to work without any *training*. It is also not unusual to find Security Personnel with several years of service who have never had the first bit of formal *training*.

Many individuals make the grade and become good Security Employees *in spite of this lack of training*.

It is perhaps the lack of *training* that causes the tremendous turnover in Security Personnel — among the highest of all professions.

The world is becoming increasingly more sophisticated and complicated and Security Personnel of the future must be better *trained*. The criminal element is constantly studying ways to outwit the law. Security Personnel must demand quality training.

Each person who reads this book is better prepared to prevent crime than most Security Persons. The author has a part in *training* personnel through *training* seminars made available to security employers by Security Seminars of Ocala, Florida.

PREVENTABLE LOSSES

Many losses suffered by individuals, businesses, industry, shipping and manufacturing could be slowed and/or stopped by the use of proper security measures and *trained* Security Personnel. A list of these losses would include:

Theft	Robbery	Burglary
Fraud	Embezzlement	Shoplifting
Hijacking	Pilferage	Records Changes
Sabotage	Malicious Destruction	

Properly designed security measures and *trained* Security Personnel can, and will, drastically reduce losses from:

Fire	Labor Violence	Vandalism
Civil Disturbances	Bombs/ Bomb Threats	Terrorism

When properly *trained* Security Personnel use security measures designed for the purpose, losses are reduced, and often prevent industrial disasters resulting from:

Explosions	Structural Collapse	Fire
Floods	Major Accidents	
Radiation Incidents		

The Security profession is one that stretches the imagination. It is very important because it has the welfare and security of every person in every walk of life dependent in whole, or in part, upon it.

And, the Security Person is a Very Important Professional. Upon his or her shoulders the safety and welfare of untold millions rest.

EMPLOYERS OF SECURITY PERSONNEL

To list all the possible employers of Security Personnel along with a list of who should be employing them would fill many pages of this book. However, there is a need for security everywhere you look. The following list of employers begins with A and ends with Z. There are hundreds more.

Airports — Agri-business — Auditoriums — Art Galleries — Athletic Fields — Apartment Complexes — Animal Farms — Architectural Firms — Auto Dealers — Bands — Bakeries — Bowling Alleys — Bus Stations — Brokers — Builders — Colleges — Credit Unions — Coin Shops — Churches — Cemeteries — Clinics — Carnivals — Department Stores — Dance Halls — Drugstores — Dairies — Discount Houses — Dental Labs — Exhibits — Engineers — Employment Agencies — Electricians — Electrical Companies — Equipment Dealers — Fruit Stands — Furniture Stores — Fairgrounds — Funeral Homes — Festivals — Florists — Grocery Stores — Golf Courses — Garages — Grain Merchants — Greenhouses — Hospitals — Hotels — Housing Projects — Hardwares — Historical Sites — Industrial Plants —Individuals — Iron Works- — Insurance Companies — Jewelry Stores — Junk Yards — Janitorial Services — Kennels — Knitting Mills — Libraries — Landowners — Logging Camps — Lawyers — Leasing Services — Landscapers — Museums — Merchandise Shows — Merchants — Motion Picture Houses — National Parks — Nurseries — Nuclear Plants — Nursing Homes — Neighborhoods — Office Buildings — Oil Companies — Observatories — Pharmacies —Painters — Quarries — Quays — Quarantine Services — Race Tracks — Radio Stations — Riding Stables — Rest Homes — Resorts — Rental Agencies — Recording Stations — Saving & Loan Associations — Schools — Synagogues — Skating

Rinks — Supermarkets — Security Guard Services — Sod Farms — Shipyards — Television Stations — Telephone Companies — Theaters — Tire Dealers — Ticket Services — Tool & Die Companies — Universities — Utility Companies — Used Car Lots — Veterinarians — Vendors — Warehouses — Weapons Centers — Wash Rooms — X-Ray Rooms — YMCA'S — YWCA'S — Zoo's.

CHAPTER II

THE JOB

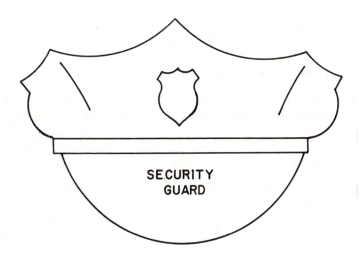

SECURITY
GUARD

THREE KINDS OF SECURITY EMPLOYMENT

Individuals desiring to enter the security profession as Security Personnel have several paths to follow into the job market. They are summarized here to show the main differences.

Proprietary Employment

An individual employed by a person, a business owner, a plant manager, a school, or any other entity, to work directly for them in the security field is said to have proprietary employment. When so employed, the individual owes all allegiance to that employer and works under that employer's supervision and is a part of the personnel staff. He reports directly to the employer and looks to that employer for training, guidance and payment. If necessary, he wears a uniform as instructed. Employment by a school board to patrol school grounds at night is proprietary.

Contract Employment

An individual employed by an agency to work for the agency in the security field is said to be a contract employee. The agency contracts with business, industry, and others to perform a number of security duties. The individual working for a contract agency owes allegiance first to the agency for whom employed and secondarily to the end user of the contract service. He reports to the agency and looks to the agency for training, guidance and payment. If necessary he wears the agency uniform and may be given many different assignments. Employment by any agency that contracts for its services to others is contract employment.

Self Employment

An individual who sells services in the field of security to others on a fee basis is said to be self employed. This person may hire out to several others at one time as long as all are properly served and there is no conflict of interest. (Working as a neighborhood night patrolman for a citizens group on a fee basis and working in crowd control at a pro football game on Sunday afternoon for a fee would be OK.)

Another form of self employment is the individual who does private investigations on a fee basis.

SECURITY PERSONNEL ON DUTY IN THE AREA OF PREVENTION:

1. Uses common sense — remains alert — uses five senses at all times.
2. Closes open windows — checks safes and file cabinets — closes doors.

3. Turns lights on or off — checks locks and latches — locks doors.
4. Disconnects electrical appliances left on — coffee pots, etc.
5. Keeps fire away from inflammable materials — drapes, couches, etc.
6. Marks danger spots — open ditches — parked trucks, rail sidings.
7. Ensures that all visitors have proper reasons to be on premises.
8. Checks outgoing shipments and traffic.
9. Keeps an eye on time clock check-ins and check-outs.
10. Has an active package control center.
11. Spot checks employee and vendor lunch boxes, brief cases, purses.
12. Installs a pilferage control system.
13. Works with inventory control to spot shortages.
14. Keeps an eye on trash and garbage removal methods.
15. Reports *all* safety violations.
16. Clears out bugging devices.
17. Periodically checks and inspects motor vehicles and trucks.
18. Controls the entrances and exits from the perimeter barrier.
19. Prevents unauthorized entrance to computer stations.
20. Checks payroll printouts for overpayments and payroll doctoring.

There are many other ways to avoid trouble that might be added to this list. However, Security Personnel must know the wishes and policies of the employer and carry them out to the letter. It is well to remember the old adage, "An ounce of prevention is worth a pound of cure." It has a great place in security work. It is easier to keep it from happening than it is to try and handle it after it has happened.

IN THE AREA OF PROTECTION:

1. Has complete knowledge of emergency orders and how to use them.
2. Has the knowledge of, and ability to operate, all protective devices.
3. Knows telephone numbers of everyone to be notified in an emergency.
4. Knows where all fire extinguishers are located.
5. Knows what kind of fire extinguishers to use on a specific type of fire.
6. Knows who can help when help is needed.
7. Knows where to go or call for help when help is needed.
8. Is fully aware of authority carried according to the law.
9. Is fully aware of the authority granted by the employer.
10. Exercises the authority granted by the law and the employer.
11. Maintains a manner that gives others confidence.
12. Keeps composure at all times. Protects others by keeping them calm.
13. Knows the limits of personal, mental, and physical capabilities.
14. Is prepared to take command in an emergency.
15. Knows all entrances and exits in the boundaries of the property.
16. Knows location of all electrical control panels, and how to turn off/on.
17. Knows location of all liquid and gas control valves and how to turn on/off.
18. Knows controls of and how to handle keyed elevators.

19. Has plans to handle unforeseen hazards.
20. Relinquishes authority, but assists professional help, when it arrives.

Security Personnel can add more items to the list of things to be done when protecting people and property. The rules and regulations of the employer are to be obeyed, as long as they are legal. Having noted the amount of knowledge the protector must have, it may pay to post the reminder, "A Little Knowledge Is A Dangerous Thing." Like the Boy Scouts, they must "Be Prepared."

IN THE AREA OF GUARDING:

1. Requires detailed instructions as to duties expected on the shift.
2. Understands the duties assigned the shift.
3. Understands the overall operations of the detecting systems.
4. Checks directions for use, and properly uses, any detecting system on hand.
5. Investigates out of the ordinary noises, odors, and sight studies the lay of the land.
6. Immediately reports any abnormal happening.
7. During patrol, frequently stops, looks and listens.
8. On patrol keeps eyes moving continuously.
9. On fixed post assignment, stays alert.
10. Spots, then gives particular attention to vulnerable areas.
11. Controls vehicular travel within boundaries of the area.
12. Checks foot traffic within boundaries of the area.
13. Approaches entrances and exits with caution.
14. Keeps safe distances from roof tops when on patrol.
15. Makes sure flashlight, nightstick (and gun, if armed), are in working condition.

16. Checks with control center before leaving on patrol.
17. Checks with control when returning from patrol.
18. Stays in touch by radio, if available, with control center.
19. Calls for assistance as soon as it is needed.
20. Makes rounds within prescribed time limits.

There are many other items that might be added to this list. The policies and instructions of the employer are to be carried out. It is well to remember an aged limerick, "Those who run away, live to guard another day, but those who stay and are slain, get not the chance to guard again." Security Personnel are careful — they use their heads. They do not become dead heroes.

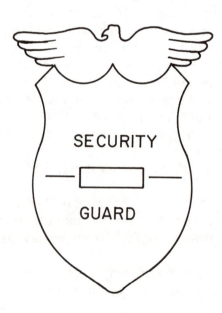

SECURITY

GUARD

SECURITY PERSONNEL ON PATROL

The duties of the Security Personnel on patrol are numerous and require close attention to details. While on patrol, Security Personnel must be ever alert — those who would harm persons and property are always seeking ways and means to do so.

Security Personnel are the first line of defense against fire. Knowing the location of fire extinguishers, knowing the type of fire extinguisher to use against each class of fire and knowing how to use them is a duty of Security Personnel. Many times quick and efficient use of on-hand fire extinguishers have prevented major fires.

Security Personnel on patrol check on housekeeping matters to keep them from causing fires. Faulty wiring, leaky flues, and areas subject to fire from spontaneous combustion are observed and reported by Security Personnel.

The security system in use on the guarded premises, whether it consists of a single Security Person or is an electronic system, must function. The responsibility to see that the system works is that of Security Personnel.

Security Personnel routinely check the perimeters or boundaries of the property patrolled. This prevents a breach of security due to open doors, open gates, open windows and/or accidental or malicious damage to fences or other boundary barriers.

During patrol, Security Personnel check locks for signs of tampering and to see if they have been left unlocked. If left unlocked, a report is made and the possibility of "inside" efforts to commit a crime is explored. No lock is burglar proof, but the deterrent feature is entirely lost if the lock is left open or damaged.

As rounds are made by Security Personnel, notice is taken of conditions near the perimeter barriers. Materials stacked near fences or walls may have been put there to aid entry to the premises. They afford an opportunity for entry that should not be present. Weeds and shrubbery near boundaries pose a problem since they afford cover for a lawbreaker. Damage to parking lots from erosion and breach of the perimeter caused by water washing under fences and walls during heavy rainfall must be investigated and reported.

Alert Security Personnel check outside areas for spots that might be used to hide trespassers. Lumber piles, stacks of pipes, and trash bins have been used by thieves and pilferers as hiding places for merchandise to be removed later. Items and materials that have been moved and placed in areas without apparent reason may be an effort by someone to get them into position for theft at a later time.

Security Personnel have the responsibility to read the temperatures of refrigerator truck interiors on the lot, to see if they have been vandalized, and to make sure they are properly secured.

The reporting of burned out lights, broken bulbs, checking lights over entrances and exits and stairways is part of patrol duty. Checking to see if damages were caused by vandalism or criminal act is part of patrol duty. Reporting damages to security headquarters and to maintenance personnel should be done as soon as possible.

Security Personnel make regular tests of mechanical alarm systems to assure they are in working order. A faulty smoke alarm, an electronic eye that fails or an alarm bell that doesn't ring is a hazard to security — and a hazard to Security Personnel.

Sharp-eyed Security Personnel are often the first to uncover employee pilferage by detecting cartons that have been broken into or that have been dropped on purpose and shoved into a corner for later removal.

On each visit to an area during the shift patrol,

Security Personnel must check computer centers, security cages and other sensitive areas. They must make sure the area is properly secured and that non-authorized and unauthorized persons are kept from the area.

Using all senses during patrol, Security Personnel stop, look, and listen at frequent intervals. Hearing escaping gas or water from mains, air handlers clattering, or a cry for help, may prevent a major disaster.

Security Personnel know which telephone lines are connected to the outside. A list of all important telephone numbers is carried on patrol.

Security Personnel check all empty spaces. The space above and the space behind. When safety hazards are found, they report verbally to the next shift, and in the written report to the shift supervisor.

Security Personnel are responsible for personal gear and company furnished items used on patrol. A clean revolver, in excellent working condition, is a must for anyone armed. A flashlight in good repair, with fresh batteries, a properly cared for night stick, communication equipment such as radio or walkie-talkie and dependable transportation available are on the "must" list. It is not difficult to keep equipment in top shape. The safety of the guarded premises as well as the life of Security Personnel may depend upon the condition of this equipment.

CHECK LIST

1. Know the location of, and check fire extinguishers by types.
2. Check housekeeping for possible spontaneous combustion fire hazards.
3. Make sure the security system is working.
4. Check for breaches of the perimeter barriers such as open gates, etc.
5. Check perimeter barrier for wire damages.

6. Check locks for signs of tampering and to be sure they are locked.
7. Make sure barriers are free of unwanted growth, debris piles, etc.
8. Keep an eye on possible damage from erosion following heavy rains.
9. Check outside areas for spots where trespassers or pilferers may hide.
10. Watch for items and materials moved without apparent reason.
11. Inspect trucks, check refrigeration temperatures, secure properly.
12. Check for burned out lights, broken bulbs and report same.
13. Check lights over entrances, stairs, etc. If tampered with, report.
14. Check all alarm systems to be sure they are working.
15. Check broken cartons for possible employee pilerage.
16. Check computer centers, security cages and sensitive areas.
17. Make rounds within the time limits outlined in post orders.
18. Investigate out of ordinary sounds and smells.
19. Report anything that appears to be other than normal.
20. Know which telephone lines connect outside — have important numbers.
21. Stop, look and listen while on patrol. Check above and behind.
22. Look for safety violations and safety hazards and report them.
23. Keep unauthorized persons off premises — observe authorized personnel.
24. Keep a watch for trouble on water, electrical and gas lines.

25. Make sure revolvers, flashlights, night sticks and communications gear is in excellent working order.
26. Check all doors and windows — remember — it was a Security Guard that discovered the break in at Watergate.

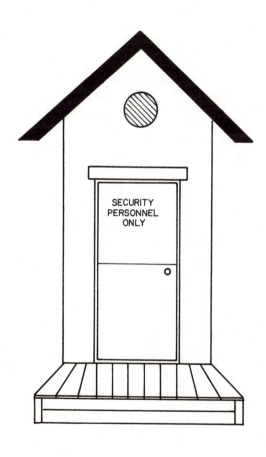

CHAPTER III

THE PROFESSIONAL

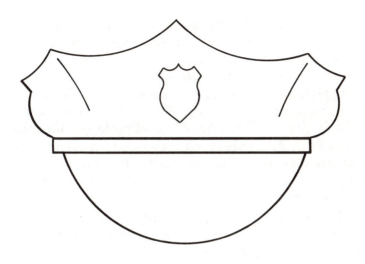

GENERAL REQUIREMENTS

There are many qualities a person must have who wishes to become employed in the field of Security. Good health, steady nerves, and a desire to serve others are among the qualities.

Security Personnel applicants must be functionally literate and able to read and write fairly well. Many assignments will require post orders that are sometimes hard to understand. Being able to read and understand is necessary. All shifts require reporting of events that occur during the shift. Security Personnel must be able to write about whatever happened during the duty hours. It doesn't have to be perfect, or in correct English, but it must be written so the details are understood by the reader.

Good health is a relative term. To Security Personnel it means physically able to perform any and all assigned tasks. Included in the physical requirements would be: weight in proportion to height, vision corrected to ordinary reading material, no disabling impairments, and above all, mental alertness.

Applicants for Security Personnel positions must not have a criminal record. They must have a current driver's license. They must be eligible for licensing by the State and able to obtain a bond. In many states a written examination, given by the State, is a requirement. It must be taken, and passed, prior to employment.

The last general requirement Security Personnel applicants must have is to be free of problems that stem from drug use, gambling or use of alcohol.

SPECIFIC REQUIREMENTS

a.   Functionally literate — able to read and write fairly well.
b.   Weight in proportion to height.
c.   Physically able to perform assigned tasks.
d.   Preferred ages — 20 to 50.
e.   Vision corrected to ordinary reading material.
f.   No criminal record.
g.   Licensed driver.
h.   Bondable.
i.   Be able to pass a written examination.
j.   Be eligible to be licensed by the State.
k.   Free of problems related to drugs —
     gambling — alcohol.

The mental attitude of Security Personnel is more important than the physical requirements.

Security Personnel who have and show confidence in themselves, who can face up to discouragement and disappointment, and come back for more of the same are Security Personnel with correct mental attitudes.

GOOD SECURITY PERSONNEL
MUST BE CAPABLE OF:

1. Performing a great number and variety of tasks.
2. Assuming a great deal of responsibility with little authority.
3. Qualifying on the range and properly handling a firearm.
4. When authorized, carrying out seizure, search and arrest duties.
5. Becoming deputized and carrying out law enforcement duties.
6. Acting to preserve life and property at great personal risk.
7. Giving protection against thieves, robbers and vandals.
8. Working with little recognition or thanks for a job well done.

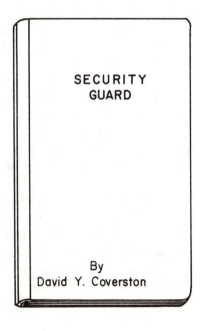

SECURITY
GUARD

By
David Y. Coverston

9. Preventing fellow employees from damaging themselves.
10. Working from fixed post or roving patrol — vehicle or on foot.
11. Investigating and reporting a wide variety of events.
12. Working closely with management and fellow officers.
13. Using ego, empathy, compassion and fairness to all people.
14. Using all five natural senses and developing a sixth.
15. Being happy with work accomplished but never satisfied.

EDUCATION AND TRAINING REQUIREMENTS

A basic requirement for Security Personnel is the ability to read and understand standing orders and special orders for the shift. Some orders are written in language that is hard to read. Some written orders are complicated. Good reading skills are needed.

At the end of the shift, a written report is required. It doesn't have to be fancy or grammatically pure, but it must be written so the reader knows what took place during the shift. The written report is often a vital defense of actions taken during the shift by Security Personnel.

Security Personnel are found with every imaginable level of education. Some have few years of formal schooling, others have fulfilled the requirements for a PhD. Successful Security Personnel are those who use the education they have to continue the learning process on a daily basis. Persons who use their educational background with common sense become winners.

PERSONAL APPEARANCE

Security Personnel may be uniformed or may be in plain clothes. In either case, their appearance should be tops. In the minds of the public, personal appearance shows ability, and Security Personnel are public figures. There is no excuse for Security Personnel being other than personally clean and neat.

Security Personnel in uniform must realize many people look upon uniformed persons with different eyes than the eyes they use in looking at others. Some looks may be a reinforcement of the respect due Security Personnel. Some looks, however, may be a rejection of the authority the uniform represents. Security Personnel must learn to deal with this fact of life.

PERSONAL RULES TO FOLLOW

1. Be bathed and deodorant clean every day.
2. Keep uniforms or plain clothes well mended, pressed or ironed.
3. Keep hair neatly combed. Mustaches and beards neatly trimmed.
4. Keep fingernails clean and trimmed.
5. Be freshly shaved. Teeth brushed and flossed.
6. Keep leather (shoes, belts) and metal (badges, insignia) shining.
7. Wear fresh underwear and shirt daily.
8. Be sure shoulder patches, name tags, and other insignia are properly attached.
9. Wear a smiling face and friendly attitude.
10. Remember, an erect figure and businesslike demeanor inspire respect.

When Security Personnel have completed work for the day, they should change from uniform to civilian clothes. The uniform is a badge of authority, and upon leaving the place of employment, that authority no longer exists.

It is understood, of course, that wearing a uniform to and from the workplace is a necessity in most cases. No one is going to question the stop to pick up some doughnuts going to work or the stop to get a loaf of bread on the way home. However, Security Personnel never wear the uniform when entering a place that could bring doubt to the mind of an observer. A bar, race track, poker club or Jai Alai fronton is no place for off-duty Security Personnel to be wearing uniforms.

GENERAL CONDUCT AND DEMEANOR

The conduct and demeanor of Security Personnel is important. Security Personnel are public figures and must present an image of one who is dependable and trustworthy.

Security Personnel must set an example of discretion coupled with authority. A good attitude towards life in general is an asset. Cockiness is out of place, but a confident and sure manner earns respect for the individual.

The Security Profession is a people business for the most part. An outgoing personality makes the Security job easier to handle. Courtesy is a contagious habit. Security Personnel who use the magic words of "please", "thank you", "Sir" and "Ma'am" are a huge step ahead of those who do not use these magic words.

Security Personnel must be honest with everyone — and that includes themselves. Telling the truth, regardless of the consequences, is the only way to survive.

Security Personnel must have pride in themselves and in the profession. The world knows a proud protector by the way the uniform is worn — clean, pressed, insignia in place, shoes and metal polished, leather in good order.

Security Personnel must be well schooled in self discipline, and have a sense of duty towards others.

Security Personnel must have integrity, loyalty and high moral character. Security Personnel do not gossip.

Security Personnel know the rules and regulations and obey them. Assignments are accepted with grace, and orders from superiors and employers are followed as long as they are legal.

Security Personnel deal with others on a firm and fair basis. Heresay is not repeated. Although cheerful towards all, they avoid being overly cordial. The rules are enforced with equal treatment for all.

Security Personnel work with other personnel in a friendly and helpful manner. On the job they understand it is better to be an hour early than a minute late.

Security Personnel take their job seriously. There is no room for gambling, goofing off, entertaining or sleeping while on the shift. The telephone is used in a proper and polished manner. In dealing with the public, the boss, the family and co-workers, each is handled with skill and tact.

Security Personnel use sound judgement, think ahead and take the initiative when required. Security Personnel must have, and use, large portions of Common Sense.

PROFESSIONAL CONDUCT
AND DEMEANOR TRAITS

1. A good and confident attitude. Not cocky, but sure.
2. An outgoing personality — Security is a people business.
3. Honesty. With self and with others.
4. To be dressed cleanly and neatly. Uniforms pressed and shoes shined.
5. To use common sense at all times.
6. A sense of duty towards others.

7. Self discipline.
8. Punctuality. Better an hour early than a minute late.
9. To be truthful. Even if it hurts, once out, penalties are less severe.
10. Discretion. Gossip and heresay are not part of the job.
11. Pride. Look like a professional — act like a professional.
12. Courteousness. "Sir", "Ma'am", "Please", "Thanks", never hurt.
13. Deportment. Conduct befitting a lady or gentleman. Others notice.
14. When in uniform, properly uniformed. Insignia and all.
15. Willing acceptance of assignments. If impossible, say so, and why.
16. Loyalty. To self, to employer, fellow employees and friends.
17. Seriousness. Gambling, sleeping and entertaining are not done on shift.
18. High morals. Never get into an awkward position.
19. Obedience. To employers and superiors. If not legal, report same.
20. Knowledge of the rules and regulations, how to obey and enforce.
21. Cheerfulness towards all — cordiality towards none.
22. Awareness that alcohol and drug use is not tolerated.
23. Knowledge of proper telephone use. Polished manners are demanded.
24. Motivation. Most of this must come from within.
25. Sound judgement. Thinking is the way to make wise decisions.
26. Firmness. A steady application of the rules towards all.

27. A sense of responsibility. Those who do not have — fail.
28. Initiative. Taking it makes you somebody.
29. Tact. When dealing with anyone.
30. Perseverance. Hang in there — don't give in. Remember the words of Winston Churchill. He said, to succeed, you must
   "NEVER, NEVER, NEVER — GIVE UP —
   NEVER, NEVER , NEVER."

MOTIVATION

Security Personnel must seek motivation from within for the most part. However, other factors favorably influence motivation. Better pay — good assignments to balance poor assignments — good fringe benefits — and increased opportunities for education and training. These are top motivational factors.

An often overlooked, but excellent motivator, is a pat on the back from the employer for a job well done. Another is public recognition of Security Personnel as workers who make this a safer and more secure world in which to rear a family.

The awarding of plaques, medals, certificates of merit, and good press coverage for exceptional work are other means of motivating Security Personnel to higher performance levels.

RESPONSIBILITIES

Security Personnel have more responsibility, less authority, and less pay than any other employee. When taking over whatever shift has been assigned, a tremendous responsibility is assumed. What other employee is given full responsibility for a million dollar business eight hours every night? What other employee has the responsibility to be the eye, ear, nose, voice and brain for thousands of workers in a multi-storied building as they pursue life, liberty and security? What other employee is responsible for the safety of a yacht basin and millions of dollars worth of boats? The answer, NONE! Only Personnel assigned to Security are given these awesome responsibilities.

Security Personnel have the responsibility to know, to observe, and to obey all rules and regulations in the work area.

Security Personnel have the responsibility to protect property — personal and otherwise — of the employer. This applies to any employer — proprietary or contract.

Security Personnel have the responsibility to obey the employer. Security Personnel have the responsibility to assist police and other law enforcement personnel in preserving the peace and in the performance of their duties.

Security Personnel have the responsibility to make legal arrest, when on duty at the duty station, as permitted by the laws of detention and arrest. They must notify police of such arrest, and surrender the suspect to them.

Security Personnel have the responsibility to carry out special assignments given by the employer.

Security Personnel have the responsibility to prevent crime, if possible, to respond to calls when they occur, and to report accidents — minor and major.

Security Personnel have the responsibility to conduct patrols in the area to which they have been assigned.

Security Personnel have the responsibility for making reports on personal activities and other activities observed while on duty.

CHAPTER IV

LEGAL GUIDELINES

SECURITY LAWS AND LEGAL ACTIONS

Except for commissioned or deputized situations, Security Personnel in many states have no more authority to make arrests or searches than any private citizen. However, when on duty for an employer, the duties may include searching, detaining and arresting individuals. If someone is observed taking property, or acting against the employer a search and/or arrest may be in order. Sometimes searches are required but they can only be performed when the employees and visitors have been previously advised.

In order to make other legal arrests, Security Personnel must know a felony was committed or witness it happening. *Felony* is a key word.

Security Personnel may not make an arrest for a *Misdemeanor* which has not been committed in their presence.

DETENTION

Permissible if probable cause to believe that merchandise or farm produce has been unlawfully taken by a person and security believes it can be recovered by holding the person for a reasonable period of time. A law enforcement officer shall be called to the scene immediately.

AND

Activation of anti-shoplifting or inventory control devices by a person leaving a protected area or business is reasonable cause for holding the person leaving if posted evidence is present to advise customers such a device is being used.

No criminal or civil liability is incurred by Security Personnel, or employer, if the above rules are followed.

A person resisting reasonable efforts of detention by Security Personnel who have probable cause to believe the person has concealed, or removed from display, or elsewhere, an item, and if found guilty of theft, is guilty of first degree misdemeanor, unless the person did not know, or have reason to know, the recovery agents were Security Personnel.

ARRESTS BY SECURITY PERSONNEL

An arrest is the detention of a person against the will when there is intent of the person arresting to detain.

Technically, an arrest is made when arresting persons identify themselves, make a statement such as "You Are Under Arrest," and either touch the suspect or the suspect agrees.

An arrest has been made when subject is under control.

The intent of the arrest must be known by the suspect.

The authority for the arrest must be known by the suspect.

Detaining a person is a technical arrest.

As soon as possible, after detaining a suspect, the police should be notified and given the details of the detention and arrest.

Use all necessary but reasonable force to effect a lawful arrest or to re-take an evading suspect.

TYPES OF ARRESTS

On View — Make the arrest after seeing the offense committed or attempted. *See it happen!*

On Complaint — Use judgement as to the validity of the complaint (complainer) and be sure to get a statement from the complaintant afterwards.

On Reasonable Suspicion of Probable Cause
This is a most troublesome problem — not enough evidence for certain conviction — but enough evidence to arouse suspicion in the mind of a reasonable person.

FALSE ARREST

Unlawful physical restraint by one person of the liberty of another person. (Security Personnel have some protection in this situation.) *One must be sure* they are arresting for acts that are contrary to law or regulation *within limits of their authority.*

RESISTING ARREST

The hindering, obstructing, pulling away, running away, striking of the arresting person. An act opposing a legal arrest by a duly constituted officer in discharging duties. Can be active, as above, or it can be passive (such as lying down and refusing to move).

HOT PURSUIT

When a person has committed a felony or misdemeanor in the presence of authority, or if an authority has reasonable suspicion that a felony or misdemeanor has been committed and the person about to be arrested committed it, the authority may, if necessary, chase the suspected person in order to arrest them, even if in another jurisdiction. This is the legal definition of "Hot Pursuit."

Security Personnel should be familiar with the meaning of hot pursuit and should know exactly what the employer expects them to do when such a situation comes up. They should know if they are expected to go in hot pursuit and to what extent.

MIRANDA RULES

Security Personnel seldom have to worry about the Miranda rules that came as the result of a ruling by the U.S. Supreme Court in 1966. The rules regard the rights of one arrested to remain silent until they can obtain the services of third party representation. The warning, usually given by the law enforcement officer who takes the person into custody, has four lines:

1. You have the right to remain silent.
2. Anything you say can and will be used against you in a court of law.
3. You have the right to consult with a lawyer and to have him/her with you while you are being questioned.
4. If you cannot afford to hire a lawyer, one will be appointed for you before any questioning, if you so desire.

As in the case of hot pursuit, Security Personnel should know what their employer wishes done if the situation arises. If questioning is to be done, the warning should be given, witnesses to the warning noted, and a statement from the person arrested and questioned that the warning was given. If the arrested person will not sign, make sure the witnesses sign such a statement.

FOURTH AMENDMENT

"The right of the people to be secure in their persons, houses, papers, and effects, against unreasonable searches and seizures, shall not be violated and no warrants shall issue, but upon probable cause, supported by oath or affirmation, and particularly describing the place to be searched, and the person or things to be seized."

SEARCHES AND SEIZURES

Security Personnel must bear in mind the Fourth Amendment when approaching a situation which requires a search or seizure of property.

There are four times when a person, an automobile or premises may be legally searched:

1. When a search warrant has been obtained. The warrant must describe accurately the place, person or auto to be searched. Any property to be seized must also be accurately described.
2. When a person consents to have person and property searched. (Getting a consent in writing is always a good idea.)
3. When a lawful arrest has been made, it is lawful to search the person, clothing worn, and property in immediate control of the person. Search cannot be made of the premises. If the arrested person is

female it is a good idea to have another female search her or make sure another female is present at the search.

4. When an emergency situation is present. The situation must be one in which immediate action is needed to prevent the removal, disposal or destruction of property reasonably believed to be goods pertaining to the crime. Examples would be the flushing down the toilet a supply of drugs, the burning of counterfeit money, or throwing away small articles such as rings or watches.

SECURITY PERSONNEL MUST REMEMBER

1. The laws of the political subdivision they work in.
2. The rules and regulations of their employers.
3. The written instructions of their post.
4. That any incident in which they are involved dealing with people and property will more than likely have an aftermath.
5. The keeping of full and accurate reports is one of the strongest defenses Security Personnel have when it comes to defending their actions.

PREPARATION FOR PROSECUTION

— Go on the assumption a prosecution will follow an incident and begin preparations at once. Remember — the accused is going on trial.

— Developed facts, and a preserved record of them, may serve as a defense against a civil suit by the accused. They are also needed by the insurance company.

APPEARING AS WITNESS

Security Personnel in court must look and act like professionals. Uniforms must be freshly and neatly pressed, leather, brass and shoes shined, insignia in place, freshly laundered shirt, and shoulder patch properly displayed.

If in plain clothes, the suit must be freshly and neatly pressed, the shirt freshly laundered, tie correctly tied, shoes shined and pockets flat.

Whatever the dress — personal grooming should be above reproach. Hair, mustaches and beards should be neatly trimmed and combed. Security Personnel must remain calm and collected.

Remember, it is the criminal on trial, not the Security Person. (It is a sure bet that the accused lawbreaker is going to be neatly dressed, well groomed, and on the best of behavior.)

TESTIFYING IN COURT

When asked a question, reply with a direct look at the jury, if a jury trial, or at the judge. All answers should be in a strong and firm voice. Always use, "Yes Sir" or "Yes Ma'am" or "No Sir" or "No Ma'am" when replying to other than the judge. In replying to the judge, address the person as "Your Honor."

Sit erect and show complete confidence in testimony. Answer all questions as briefly as possible. Do not be afraid to say "I do not know the answer to that question." Do not be afraid to say "I cannot recall that event."

The prosecuting attorney will have gone over the testimony to be given and that is the testimony that must be given. If asked by the defense attorney, the judge, or other proper authority, if the case has been discussed, the reply should be "Yes, with the prosecuting attorney." If asked if the prosecuting attorney advised what to say, the reply should be, "No, I speak for myself."

The attorney for the defendant will try to trip you up and will try to put words in your mouth. Do not let that attorney bait you. Listen closely to all questions and if you are not clear as to the question, ask for it to be repeated. Think about the answer, reply slowly, and with confidence. The defense attorney will do anything within the law to save the defendant. It helps that cause if you can be confused, caused to panic, lose your temper, or otherwise lose credibility.

Reply to all questions directed to you firmly and with certainty. Do not use the words, "I think" —you either know or do not know. Do not lie. Do not express opinions. Do not guess. Do not volunteer additional information.

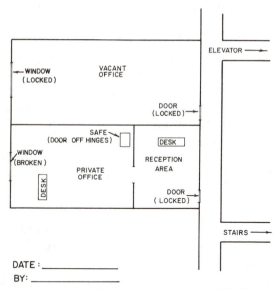

INDOOR SCENE SKETCH

2nd. FLOOR - CLIMER BUILDING

BROWN LAND SURVEYING SUITE

ELEVATOR ——▶

←WINDOW
(LOCKED)

VACANT
OFFICE

DOOR
(LOCKED)—▶

SAFE
(DOOR OFF HINGES)

DESK

WINDOW
(BROKEN)

RECEPTION
AREA

PRIVATE
OFFICE

DESK

DOOR
(LOCKED)

STAIRS ——▶

DATE :_____
BY: _____

Before you appear in court, go over all of the details of the situation in your mind. Refer to your notes, sketches and diagrams you or others may have made. Be prepared.

ABOVE EVERYTHING ELSE — TELL THE TRUTH.

SEVEN POINTS TO REMEMBER

1. Do not lie.
2. Do not guess.
3. Do not volunteer information.
4. Do not argue.
5. Do not be afraid.
6. Do not reply too quickly.
7. Do not lose temper.

CHAPTER V

CRIMINAL ACTS

TWO TYPES OF CRIMES

TYPE 1 — Offenses by people who had permission to be on the property such as employees and customers.

TYPE 2 — Offenses by people who did not have permission to be on the property, such as burglars, robbers, trespassers.

The prosecution of criminal suspects comes from the employers of Security Personnel — if a proprietary employee. From the agency if a contract employee.

PROPERTY CRIMES

General Larceny — This crime is the taking away without consent of the owner, and against the will of the owner, property that belongs to another. It is the willful and felonious intent to permanently deprive the

59

owner of property. The property could be items above certain dollar amounts (varies within the states of the U.S.) such as automobiles, money, goods, merchandise, etc. General larceny is usually classed as either petty or grand larceny. The difference between petty and grand larceny is a matter of value of the things stolen.

From A Person — This crime has the elements of general larceny. In a technical sense, the taking must have been from the person who owns the property or a taking of property while in the presence of the owner. Purse snatching is a taking from a person. Stealing a CB radio from a car while an owner is moving to stop the theft is taking of property while in the presence of the owner.

From A Building - The elements of general larceny are present but technically it is taking property from an building, a dwelling, office or other structure. Taking a typewriter from an office during business hours or while the custodial crew is cleaning up after hours is a good example.

From A Motor Vehicle - The elements of general larceny are again present but the taking must have been from the outside of the motor vehicle (stealing tires or hubcaps), or from the inside of an unlocked motor vehicle (stealing Christmas packages from an unlocked vehicle).

Note — If the theft is from the inside of a vehicle that has been locked and has been entered unlawfully, the crime is called burglary. Most insurance companies will pay for stolen contents only if the vehicle was locked and the thief had to break and enter the vehicle in some manner.

MALICIOUS DESTRUCTION OF PROPERTY

To be charged with this crime it must be shown that a person or persons caused damage or injury to the personal property or structure of another with willful and malicious intent. Examples would include: (1) pushing over a fence belonging to another with intent to release pets; (2) turning water faucets on so water runs while an owner is absent; (3) throwing rocks and breaking windows.

VANDALISM

Vandalism is the intentional damaging of private or public properties. Unlike the previous discussed *Malicious Destruction of Property* (willful and malicious intent), this serious crime is generally unexplained. The tossing of a rock through a window, slashing the tires of a hotel guest, driving a vehicle across a golf course green, smashing furniture and breaking shrubs are examples. Although this crime is usually committed by juveniles, it is also committed by adults. To help prevent this crime, Security Personnel should be in uniform, be highly visible and patrol irregular routes.

SABOTAGE

Sabotage takes a great number of forms. It can be sabotage by starting fires, exploding bombs, opening water valves, destructing scale models, wrecking machinery or cutting power and telephone lines. Anything done to slow progress is a form of sabotage.

REASONS FOR SABOTAGE

A. Social protest. A fellow worker has been fired or the company has announced a cutback in benefits.
B. Revenge. Real or imagined, this is a powerful motive. A fired worker or the denial of a claim may trigger revenge.
C. To injure a particular property. A classic example is the kid who burns down the school house.

D. Spite. The bank wouldn't honor a check drawn on an out of town bank so a pipe bomb is dropped in the night depository.
E. Military slowdown. This form of sabotage is the romantic type shown in motion pictures. It is practiced by both sides in war. The blowing up of bridges, the loosening of railroad tracks, etc.

TERRORISM

Terrorists continue to make headlines the world over. Touching on the problem here is necessary. Many Security Officers who read this will continue upward in the field of security and upon reaching high positions may be faced with the possibility of dealing with the terrorist. The study of terrorists and terrorism is a complete study in itself.

TERRORIST ACTIVITIES

1. Bombings.
2. Kidnapping and Taking of Hostages.
3. Assassinations.
4. Hijackings.
5. Attacks Upon and Seizure of Facilities.

THE CLASSES OF TERRORISTS INCLUDE:

A. Political classes. (The P.L.O. and Patty Hearst situations.)
B. Minorities. (The Black Liberation Army.)
C. Criminal-Money oriented. (The Italian Moro case.)
D. Mentally Disturbed-Imagined wrongs. (The Washington Monument case.)
E. Religious. (The Black Muslims.)

ASSASSINATIONS

An individual has no value when dead. For this reason, it is usually a loner or a planned effort to do away with a person or persons that leds to an assassination. The true meaning of assassination is to kill the head of a state, such as Lincoln, Garfield McKinley, Kennedy, or Sadat. However, in recent years it has taken on the meaning of the murder of a prominent person. Huey P. Long and Robert Kennedy were United States Senators when they were murdered and the term "assassinated" was used in both situations.

It must be remembered — assassins will attack anyone who gets in their way. In the attempt on the life of President Reagan, press secretary James Brady was seriously wounded.

TRESPASS

The word has many meanings. To Security Personnel. it means one thing — *To Enter Onto Another's Property Illegally.*

Trespassers and other intruders pose a threat to security because it is hard to know what the trespass or intrusion is about. It could be someone who is lost and seeking help. It could also be someone who is running from a criminal act. Or, it could be someone who is testing the quality of the security measures in place.

Whatever the reason — only the trespasser knows — and Security Personnel must handle the situation with care. The person, or persons, may be desperate and criminal in nature, psychopathic, or burglary minded.

Never assume a trespasser to be other than someone to be treated with caution and one who is up to no good. Never assume a trespasser or trespassers to be alone. Never allow an opening to attack by trespassers.

When on patrol, keep the thinking cap on at all times. Doors to be entered may have intruders behind them, ready to pounce. At night the flashlight carried may be taken away. On night patrol the careful Security Person keeps back light from making the patroller a target for a trespasser. When patrolling building interiors at night Security Personnel should slam doors against the wall as they pass through. They should keep their flashlight on thumb pressure only so it goes out if accidentally dropped or is knocked from the hand.

Security Personnel must continually remind themselves the criminal mind is never asleep. Criminals are out to do damage at a risk to themselves but they try to make the odds heavy in their favor. The criminal is always thinking of new ways to commit mischief so Security Personnel must ever be thinking of ways to beat them at the game.

CHAPTER VI

HAZARDOUS HUMANS

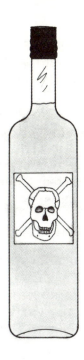

DRUNKS, DRUG USERS, AND
MENTAL INCOMPETENTS

In the society of today, there are more and more people who turn to alcohol and other drugs. They no longer stay home when using stimulants and depressives. They are found in many places. Security Personnel must learn to recognize symptoms shown by them. To protect the public, and themselves, they must know how to handle these users when it becomes necessary.

The drug user is a threat to the welfare of the nation and to security. The user has given up most, if not all, self control. Drug users are dangerous to themselves and to others. Many of them will do anything to support the habit. Drug dealers and drug pushers are often more dangerous than other users because they keep a low profile when they become addicted to their own poison.

Drug users show symptoms ranging from talkative and happy to aggressive and violent. Babbling and nervousness appear near the end of their "high" period. Then they need a new "fix".

Drug users do not fit any special pattern in society. They include males and females; they may be fat or thin, poor, well to do, short, tall, young or old. Their common trait is the need to continue the habit to which they have become addicted. They must be viewed as very dangerous individuals.

DRUGS USED AND THEIR SYMPTOMS

1. Sniffers of glue, gasoline, paint and typing correction fluid display a variety of symptoms. They may sneeze, cough, appear sleepy or drunk and find it hard to speak clearly. They behave poorly and tend to get violent. The habit is damaging their brains.
2. Marijuana smokers vary in behavior from happiness to sleepiness. They tend to lose coordination of limbs and mental alertness. They may hallucinate, become overly hungry, and develop a craving for sweets.
3. Morphine, Codeine, and Heroin users may suffer a loss of appetite. They may develop constricted eye functions and become drowsy and stupified.
4. Amphetamines and other pep pills called "uppers" tend to make users aggressive and exhibit anti-social behavorial patterns. They talk rapidly, eyes dilate, and they become confused and restless.
5. Barbituates, goof balls and other pills, known as "downers", cause the user to become incoherent, depressed and to appear impulsive and drunken.

ALCOHOLICS

The most widely used drug in the world is alcohol.

Excessive use of alcohol leads to alcoholism which stems from having been drunk too often.

The drunk does not follow a set pattern of behavior. Some drunks are mean and look for trouble. Some drunks are happy and wish to be friendly. Security Personnel must be cautious with either because the individual under the influence of alcohol is not in full possession of all mental faculties.

Another type of drunk is the sad drunk. This drunk is usually seeking sympathy and is ordinarily no more trouble than the happy drunk but caution is the watchword. Never take the actions of a drunken person for granted.

THE MENTALLY INCOMPETENT

A few words regarding the mentally incompetent. Not all such persons are being taken care of in institutions. A great many of them walk around in society. Their behavior is not predictable, and no chances should be taken with them.

Security Personnel are people watchers, and any behavior on the part of an individual that appears out of the ordinary should be investigated.

When nature deprives a person of one capability, that person usually develops other characteristics, to make up for the loss. Many a person with a muddled mind has an extraordinarily strong body. Be very careful.

THIEVES

Stealing is Stealing — no matter what is stolen, from whom stolen — how stolen — or where stolen.

Theft from an employer's stock is called pilferage. The items could be of great value, such as a television set or a refrigerator, or of small value, such as paper clips, pens or staplers.

A theft from employers seldom called theft is that of *time*. The employee taking time from the job to get a haircut or manicure without permission is stealing from the boss!

Theft by persons not employed by a business, during business hours, and without display of force, is ordinarily called shoplifting.

Theft by persons with a display of force is called robbery.

Theft by persons unlawfully entering a business is called burglary and the charge is "breaking and entering."

EMPLOYEE THEFT

The greatest losses in drug stores, supermarkets, etc., are from employee theft. This form of stealing company assets is difficult to control. The dishonest employee has inside knowledge of the business and is able to conceal activities. There are a number of methods available to combat this type of thief, but it requires unceasing alertness and aggressiveness by security personnel, security education for employees, and anything else the employer can use to stem the tide.

There are two distinct types of pilferers. The *Casual pilferer and the Systematic* pilferer. They are equally guilty of thievery.

The *Casual* pilferer steals as the opportunity arises. The item stolen is kept for personal use, given as a gift, or is sold quickly to raise money. This thief is usually not a risk taker, and does not steal unless confident the chances or being caught are slight. The best deterrent to this type of crime and criminal is to spot check items and keep close inventory control. *Do not forget*, the casual pilferer becomes more bold and less guarded as successes continue, and they grow into systematic pilferers.

The *Systematic* pilferer has a plan to steal on a regular basis from a protected area. This thief makes plans after the security plan being used at the work place is in effect. It takes advantage of weaknesses in the plan. This is an organized individual, or perhaps a group of individuals, within the organization. The best deterrent to this type of crime, and criminal, is close inventory control, rotation of employees, and spot checks of purses, brief cases, delivery tickets and employee automobiles.

HOW EMPLOYEES STEAL

1. Picking up an item as if it were a personal item. This may be in full view of fellow employees, such acts include placing compacts in purses, pens in pockets, file folders in brief cases, etc.
2. Placing an item in a coat that is not worn but is casually thrown over a shoulder when leaving the premises.
3. Having packages filled with stolen items, wrapped and addressed as if carrying out to be mailed.
4. In brief cases and purses.
5. Hauling out in motor vehicles. (Employees should not be allowed to visit their automobiles during the work period.)
6. Throwing items from windows or doors to be picked up later.
7. Using trash and garbage containers to conceal items.
8. Placing items in railroad cars for later pick up.
9. Loading items into delivery trucks for supposed delivery then removing, after hours.
10. Concealing items in clothing (socks, bras, top coats, raincoats, etc.)

HOW TO DISCOURAGE PILFERAGE

Make it be known that all employees are subject to searches of persons and vehicles at unannounced times and places — *and carry it out!* When an employee is caught — *prosecute and dismiss!* It takes only a few such moves to make employees understand it isn't worth the effort.

Establish an effective package, material, vehicle, and key control system and keep checking to see that it is working.

Locate employee parking outside the perimeter of the business.

Carefully screen during the hiring process — *check references.*

Have enough Security Personnel to patrol grounds. Have grounds checked on a constant basis for material out of place and hidden for later removal.

Keep strict inventory of easily stolen merchandise and control amounts that move from storage to the use area.

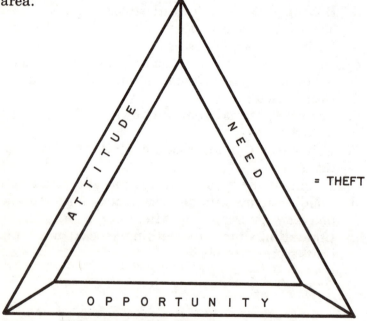

THE THREE SIDED TRIANGLE OF THEFT

Keep records of inventory and account for purchase, use, and salvage of stock and supplies.

Use a program of education to convince employees stealing is morally wrong, no matter what the value of the items taken. Let employees know it is their responsibility to report any theft.

Keep premises well lighted and fenced.

Mark all tools and equipment so they can be identified and have all borrowed equipment signed for with complete data.

ROBBERS

Robbery is the theft of something of value from another while using some form of violence. The use of guns, knives, clubs or perhaps superior muscle power to take goods or property of another, is robbery. Unarmed Security Personnel do not usually deter a robber once a robbery has begun. But, the presence of unarmed Security personnel may be the measure that prevents a robbery form taking place.

IN THE EVENT OF ROBBERY

1. Notify the law.
2. Get medical assistance if needed.
3. Secure and protect the premises.
4. Notify the security supervisor.
5. Get a physical description of the suspect or suspects.
6. Make note of the escape route taken by the suspect.
7. Interview the victim or victims and get statements if possible.
8. Get a description of the property taken.
9. Make notes to refer to in making an incident report.
10. *Above all — Do not become a dead hero!*

BURGLARS

At common law, the definition of what is known as Burglary, was the breaking into and entering of the dwelling house of another during the night with intent to steal.

For reasons *not* clear to anyone, this definition was filled with loopholes. What about breaking and entering during the day? What about breaking and entering to vandalize?

Realizing the uncertain and unclear definition given, most states have now defined criminal acts more clearly.

Statutes of today generally define Burglary, and Breaking And Entering, as one and the same. The law against Burglary defines this criminal act as "the entering of a structure (dwelling, building, mobile home, airplane, boat, etc.) with intent to steal or to commit another felony." These new Burglary statutes do away with the need for the crime classification of Breaking And Entering.

Burglary is a crime that businessmen can actively combat. If they listen to information given them and then act upon the information, they can reduce burglary considerably.

HOW TO PREVENT BURGLARY

1. Install modern locks, see they are used, and install strict key control.
2. Make sure windows, doors and skylights are secured at night.
3. Cover windows with wire mesh — take extra precautions with windows or other openings facing fire escapes and/or hidden alleys.
4. Know where all ducts, sewers and other tunnels are located, placing them on a premises map with

their intended use. If any are large enough to permit human entry, make sure they are covered at their entrance into premises. Man-safe closures such as steel bars, razor steel or the like are good materials to use.

5. Keep the interiors of buildings brightly lit. This is particularly true of entrance ways, cash handling stations, and safe locations. The burglar does not like light.

6. Keep signs and items that clutter up front windows to a minimum. Security Personnel and law enforcement officers should be able to see into the area at a glance.

7. Keep weeds and debris cleaned away from the building. Be sure your landscaping is designed with security in mind. A recent murderer blew away a girl at a bank night depository shooting from the cover of shrubbery.

8. Install a good alarm system and keep it in working order.

9. Put cash and other valuables into the safe at night. Do not leave cash in cash registers.

10. Keep a check on salespeople (they should have credentials) and other persons on or about the premises. See to it they do not have the opportunity to tamper with locks, alarms, cash registers, and safes.

11. When cash flow is expected to be heavy, sales specials, before holidays, etc., the police should be notified and additional Security Personnel assigned.

12. Security Personnel should be kept informed as to when the business is to be closed for repairs, vacations, etc.

13. All equipment should be serially numbered and records kept of them.

14. If suspicious persons are loitering near the pre-

mises and/or it is suspected the business is being "cased", Security Personnel should notify proper law enforcement agencies.

15. A detailed burglary survey should be made of the premises. When vulnerable spots are located they should be corrected immediately.

16. All prospective employees should be investigated. Many burglars apply for jobs to obtain employment in order to commit future burglary.

17. All cash registers should be left open at night. This prevents damage from prying with tools in case the burglar gets inside.

18. Blank checks and check writers should be kept in the safe.

19. Property keys should not be kept with car keys. Keys are easily duplicated in a short period of time and property keys should never leave the possession of the person accountable for them.

20. Before closing for the night, Security Personnel should check rest rooms, under counter spaces, basements, furnace rooms, electrical panel areas, custodial supply closets and all other places a person might hide. Be sure that no one remains after hours except designated personnel.

SHOPLIFTERS

The employee thief and the vendor thief are joined by a third thief who steals while the business is open. This thief is the Shoplifter. Shoplifting is one of the greatest problems facing the merchant. If all the property losses due to larceny, burglary, auto theft and robbery in North America each year were added together they would not reach the Multi-Billion Dollar losses per year from shoplifting. Shoplifting runs from 2 to 5 percent of Gross Sales.

Shoplifters come in many different sizes and colors,

with many different reasons for stealing. The shop-lifter may be young or old, male or female, and may be poor or well to do.

TYPES OF SHOPLIFTERS

1. The Professional —   This shoplifter composes about 5% of the shoplifting population. He or she steals for a living and is an expert in the field.

2. The Amateur —   This shoplifter steals for something to use personally or perhaps to give to someone else as a gift.

3. The Vagrant —   This shoplifter steals items to be used for sale or for trade.

4. The Drug Addict —   This shoplifter steals items to sell in order to get money to support the habit.

5. The Juvenile —   This shoplifter steals for a number of reasons. It may be for the thrill of trying to see if it can be done without getting caught. It may be due to peer pressure or it may be to get items to sell, to keep, or give away.

6. The Kleptomaniac —   This is a compulsive shop-lifter. The number of Klepto-maniacs is actually small.

Security Personnel should make a study of all types of shoplifters. All persons working in areas where shoplifting may occur should learn the methods of operations used by shoplifters. They should learn how to detect shoplifters and how to physically discourage this crime. Needless to say, it is Security Personnel who must first learn how to combat shoplifters, then teach it to others.

Strict inventory control by management is a prime measure to be used in determining the amount of goods being lifted by shoplifters. Management should have a strong policy concerning shoplifters made with the aid and suggestions of Security Personnel. These policies should be known by all employees, including Security Personnel. Most important - once the policy is set, it should be followed to the letter.

HOW SHOPLIFTERS ACT

1. The Professional — Cool, methodical and brazen. A self confident thief who steals in order to make a living and is very danerous.

2. The Amateur — Usually shy and nervous. Is shoplifting on an impulse of the moment.

3. The Vagrant — Usually shabbily dressed and very nervous. Is often found taking items and then running from the store.

4. The Drug Addict — Usually nervous and high strung and needing a fix, this shoplifter is a very dangerous individual.

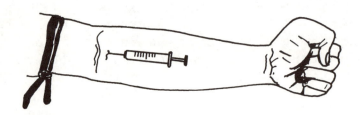

5. The Juvenile —	The majority of juvenile shoplifters are teenage females. They travel in groups and steal teen-age type merchandise.
6. The Kleptomaniac —	Usually middle aged, they appear to be above reproach. There are not as many kleptomaniacs as is ordinarily believed.

ACTIONS TAKEN BY SHOPLIFTERS

A. A shopper leaves an area of the store, or the store, in a hurry.

B. A shopper makes a number of trips to the restroom.

C. A shopper has stolen items mixed with other merchandise.

D. A shopper wearing baggy clothes — especially in warm weather.

E. A shopper makes a number of unusual motions.

F. A shopper goes behind counters.

G. A shopper is "fussy" — not interested in articles being shown.

H. A shopper appears to be more interested in watching people than in shopping for goods.

I. A shopper interchanges items while going through the store.

J. A shopper seems to be nervous and sweats a great deal.

K. A shopper keeps hand in outer coat pocket.

L. A shopper from a group of two or more engages clerk in conversation

Remember — some of the above actions may apply to anyone — but they are actions that should be watched.

SHOPLIFTING METHODS

1. Palming articles into other purchases — a comb into a shoe.
2. Using an umbrella, purse or newspaper to hide items.
3. Slit in outer garment allows articles to be held inside.
4. Wearing "shoplifter bloomers"or garments of like nature.
5. Putting on new clothes and wearing old clothes over them.
6. Use of hooks inside clothes — like a magician — to hold items
7. Wearing items out of store — entering bareheaded, leaving with hat.
8. Wearing outercoat and holding articles between legs.
9. Grabbing item from unattended area and running.
10. Starting distraction so partners can steal.
11. Switching price tags.
12. Hiding goods in baby carriages and strollers.

Security Personnel, some in uniform, some in plain clothes, with at least one armed, form the first line of defense against shoplifters. The deterrent effect of the uniform, and an armed Security Person is great. This, coupled with plain clothed Security Personnel detaining shoplifters, marks the business as one to be avoided. It is cheaper to keep shoplifting from happening than it is to catch thieves and prosecute them.

HOW TO PREVENT SHOPLIFTING

A. Service customers quickly and efficiently — shoplifters note this speed.
B. Sales people should be trained to say "Be with you in a moment."

C. Employees should be taught never to turn their back on shoppers.

D. Loiterers and wandering shoppers should be kept in sight.

E. Customers should obtain refunds from a single area.

F. No department should ever be left unattended. Thieves will pretend to be clerks if given the opportunity.

G. Develop a code warning system understood by all employees.

H. Develop simple ways for employees to alert security to shoplifters.

I. Keep expensive merchandise in view of two salespersons and keep it locked between showings.

J. Keep merchandise where it can be easily seen.

K. Arrange merchandise so it must be picked up. Two sided shelves without back dividers make it easy to stand in one isle and push items into containers in the other aisle.

L. Display only one of a pair of items — shoes, earrings, etc.

M. Keep counters and tables neat and orderly. It is hard to tell if articles are missing from jumbled displays.

N. Make merchandise hard to remove without aid of the clerk.

O. Be sure salespeople can see customers while on telephone.

P. After customer inspects an item, but does not buy it — return to stock immediately.

Q. Keep discarded sales receipts picked up and destroyed. It helps keep down illegal returns of items.

R. Open cash register only upon sale. It should be closed while merchandise is wrapped and locked when not being used.

S. All cashiers should close and lock cash drawers

whenever any kind of distraction occurs.

These have been suggestions to help prevent shoplifting. Security Personnel may have other preventative methods to add to this list.

Remember — the shoplifter is trying to figure out ways to get around security. Never give the shoplifter an opportunity.

CHAPTER VII

THE WORST CRIMINALS

BOMBS AND BOMB THREATS

Security Personnel must develop a sound attitude toward Bombs and Bomb Threats. They must be prepared to face these situations with a calm attitude and professional-like approach. The world has sick-minded people who threaten others with bombing for any number of reasons and far too many of them actually plant bombs. It is fortunate many planted bombs fail to explode. It is also fortunate many of these bombs are located and rendered harmless before they do damage.

Three common bombs, all home built, Security Personnel must learn to cope with are listed below, with their dangers:

1. **Gunpowder Bombs** — These bombs are usually low-yield bombs. Their main danger comes from their

placement and the resulting aftermath. For example, a low yield bomb that explodes in the area of stored fuels can create a major disaster.

2. **Dynamite Bombs With Timing Device** — These bombs can be very potent, according to the amount of material used in their construction, and the type of substances used. And, as is the Gunpowder Bomb, the explosion set off may trigger even greater explosions.

3. **Incendiary Bombs** — These bombs are designed to ignite fires. They are made from numerous materials. The "Molotov Cocktail" with fuel-filled glass container and lighted wick is used against automobiles. It is also tossed into buildings. Ordinary bundles of rags soaked with gasoline or oil may be set on fire and used against homes and/or other buildings.

Bomb threats are usually called in to the building in which the bomb has been placed. Since this is a common occurrence, Security Personnel should see to it that all buildings to which they have been assigned have, or will put into effect immediately, a bomb threat notice similar to the following:

NOTICE

IN CASE OF A BOMB THREAT

All telephone operators and receptionists must know exactly how to react to a bomb threat. A list at the telephone is a must. The operator or receptionist must remain calm and courteous. Get all the information possible by listening closely and not interrupting. Keep the caller on the line as long as possible by asking questions. Questions to ask the caller:

A. What time is the bomb set to go off?
B. Where is the bomb?
C. What kind of bomb is it?
D. Where are you?
E. Why are you bombing us?
F. What do you do?
G. What is your name?
H. When did you plant the bomb?
I. Are you alone?
J. Have you bombed other places?
K. Is the bomb near me?
L. What is the last place you bombed?

(A form should be completed as questions are asked of the caller. Using these, and other questions might cause a person making a bomb threat give out information, and aid in making an evaluation of the bomb threat.)

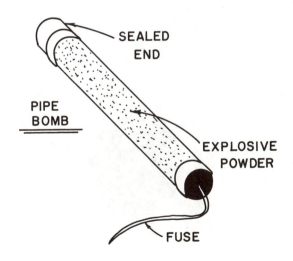

87

EVALUATION

In making an evaluation of the bomb threat, common sense should prevail. Each of the following individuals fit the profile of someone who might call in a false bomb threat: child, prankster, drunk, someone laughing or incoherent.

The telephone operator should have another person monitoring calls, if possible, and they should work together in making the first analysis of any bomb threat call. Although police and fire forces should be alerted immediately, the first decision must be made in-house.

If effective control of persons and materials entering the premises has been, and is, in effect, the lower the probability a bomb has been placed.

The more believable the warning, with details and specifics, the greater the probability a bomb has been placed.

Explosions usually damage specific areas. The areas most accessible to the public, either inside or outside, are favored spots to place bombs. The most common placement areas are the toilets, elevator shafts, under stairs, utility rooms, electrical panel locations and the heating/cooling areas.

SEARCH FOR EXPLOSIVES

If it is determined there is some certainty that a bomb has been placed, an immediate search should begin.

The best searchers are those who work in the area or who use the area the most. They are able to spot something out of the ordinary quickly. If management or Security Personnel have had "bomb drills" in the area, the job will be much easier. If not, search leaders (hopefully volunteers) are given the order to start the search.

Searching begins at a door and follows a circular path until the entire area has been searched. When the area has been searched, it should be closed off, and a notation made that it has been searched.

Searchers should check lunch buckets, ash trays, thermos bottles, waste baskets, filing cabinets, TV Monitors, Word Processors and desk drawers. Any potential hiding place is suspect until after it has been searched.

Security Personnel in charge of searches should instruct the members of their team to follow the following routine:

1. Upon entering a room, stop, search the room with the eyes, and listen for out-of-the ordinary sounds.
2. Each search party member must move slowly, first searching the area from the floor to the height of the waist. The second search is the area from waist height to the ceiling. A more intense search should be made of the area from waist height to head height, if time permits.
3. If the area to be searched is a large area, two teams should search at the same time. The teams should start their searches from opposite sides of the room and work toward the center.
4. Team members should be persons who are not subject to panic. In the process of searching in a

thorough manner, the nerves of most people settle down and a calmer atmosphere prevails.

6. No one should pick up or move any suspicious looking article. If one is found, notify the leader and clear the room.

7. If a bomb is found, notify the leader at once and clear the room. Do not make any announcement as to a finding.

8. If a bomb or suspicious article is not found during the search, report to the leader the assigned area has been searched completely and a new assignment is desired.

If a package is found that is believed to contain a bomb or other explosive device, the following precautions should be taken:

A. Do not put the package in water. Water conducts electricity and if the package gets wet, water may trigger the bomb. Also, the bomb may contain chemicals which are activated by moisture and may cause the bomb to explode.

B. Do not allow anyone to move a suspected package except those given such authority. Bomb squad people know how to move them.

C. Do not allow anyone to open a suspect package except bomb squad people.

D. Do not touch or allow any unauthorized person to touch a suspect package. Movement may trigger an explosion.

E. Do not allow anyone to smoke when making bomb searches.

Following these rules to the letter may prevent serious injury to the searchers, the building and the area. Bombers may employ sophisticated devices such as mercury switches and movement triggered devices. They may tie strings to activators, use acid-chemical combinations that explode when mixed or use other ingenious methods to achieve their goals.

Security Personnel must maintain their calm and businesslike approach to the bomb threat situation. They must move with authority and keep search team members from speaking until each task has been completed. When law enforcement personnel arrive, they should cooperate with them until the situation has cleared, then return to regular duty. A detailed report should be given supervisors as soon as possible.

EVACUATION

The three factors to be considered when making the decision to evacuate or not to evacuate a building under bomb threat are:

1. Is the bomb threat credible? The threat must be carefully, but speedily evaluated. The knowledge the threatening person has about the location of the bomb, the area it is supposedly placed in, and how accessible that area is to the general public are important things to be considered in making the evaluation. The motive, if revealed, for placing the bomb must be considered.

2. The occupancy factor. In a multi-storied building, the floor indicated as that where the bomb is planted is of main concern, with great concern for the floors immediately below and above. In a one or two-story building, the entire building is of prime concern.

3. The time factor. If short notice must be given, the immediate suspected area should be evacuated first. After the immediate area has been cleared, adjoining areas should be evacuated.

If a bomb has not been found and there is no reason to believe there was one, when should personnel who have been evacuated be allowed to return? This is a management decision.

Management should give the option of returning, or not, to the occupants. It gives them a sense of importance by being included in the decision making process. It also tends to calm people and helps prevent panic. Those who are calm will help keep others calm. Security Personnel must obey management decisions

If the operation is one that depends upon work in another area, such as an assembly plant, it is better to stop all work for the day. Security Personnel must remain on duty until their employer relieves them.

If the evacuation takes place in the morning, consideration of the aftermath must be considered. For example, the time lost to gossip about the scare. If the evacuation takes place after noon, it is probably best to send everyone home except those needed to secure the premises. Security Personnel must be extra alert when securing property after the evacuation of people has been completed.

IF A BOMB OR DEVICE EXPLODES, SECURITY PERSONNEL WILL FOLLOW FIRE AND OTHER DAMAGE CONTROL DUTIES UNTIL RELIEVED. AUTHORITIES SHOULD MAKE DECISIONS IN COOPERATION WITH MANAGEMENT, POLICE, FIRE CHIEF, AND SECURITY SUPERVISORS.

CHAPTER VIII

GUNS

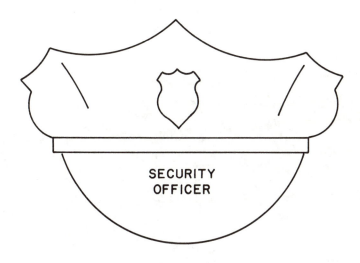

SECURITY
OFFICER

ARMED SECURITY PERSONNEL

There have been arguments, pro and con, as to whether Security Personnel should be armed. The argument has been raging for years, and will probably continue. However, many Security Persons are licensed to carry arms, and do so.

The decision to use an armed guard, or an unarmed guard, is a joint decision between the employer of the guard and the guard agency if an agency supplies the guard. If in-house Security Personnel are used, the decision is that of the employer.

The decision depends upon a number of factors.

1. Is the assignment important enough to merit the possible use of deadly force? Assignments to guard shipments of money in armored cars should merit assignment of armed guards. The assignment requiring guard duty in a hotel lobby would not.

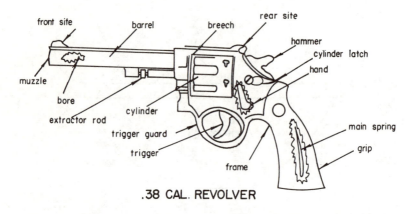

.38 CAL. REVOLVER

2. Does the assignment require Security Personnel to protect themselves, as may be the case on yard patrol at a shipyard? The decision should be yes. Guards directing traffic at a sports event need not be armed.

3. Is the potential loss worth the risk involved? If Security Personnel are assigned to cover a Presidential visit, the answer is yes. If the assignment is to prevent fishing in a public area, the answer is no.

4. Does the liability insurance carrier consider it an assignment that requires armed Security Personnel? Maybe yes, maybe no, read the policy.

5. Is the assignment in a high crime area? If so arm the guards.

Security Personnel to be armed should be selected with great care. They should have high morals, good physical ability and sound mental characteristics. Only top of the line Security Personnel should be considered for armed guard training. Armed guards should be assigned only to those areas that have been determined to need armed guard protection.

Guns to be used by Security Personnel should be provided by the employer. This precaution allows the guns to be maintained in serviceable condition and to be stored in a safe place when not being used. The same rules should apply to ammunition for the guns.

If Security Personnel are permitted to carry their own guns, the agency or other employer should require regular inspection reports on the gun from a qualified gunsmith. The report should show the gun has been inspected, is in good working condition, and is being cared for in recommended manner. This report should also indicate proper and fresh ammunition is on hand.

For the protection of employers and for the protection of Security Personnel, armed personnel must meet the following requirements, as a minimum, if they are to be used in armed guard situations.

1. Have current qualification credentials from an approved firing range.
2. Agree to annual re-qualifying at an approved firing range.
3. The gun to be carried on assignment is the same gun used on the approved firing range whether it be employer or employee owned.
4. Must understand the care and handling of firearms and keep proper and fresh ammunition.
5. Must understand the restrictions on the use of firearms and follow them to the letter.

Safety rules for armed Security Personnel are the same as the rules are for anyone else. However, it is absolutely essential for Security Personnel to know and practice the rules at all times. The key words are:

KNOW AND PRACTICE

FIREARM SAFETY RULES

1. Any gun is always to be considered a loaded gun. No exceptions. Anyone picking up a gun, or being handed a gun, *must check to see if it is loaded.* Do not allow anyone, member of the family, friend, fellow worker or supervisor to handle a gun unless they know how to check a gun to see if it is loaded, and unless they check to see if it is loaded.
2. When storing a gun, unload the gun, unless it must be ready for instant use upon removal from storage. Make sure the storage place is safe from all but authorized persons.
3. On the range, the gun must be checked each time it is handled. It must be kept muzzle pointed down range — loaded or unloaded. The gun must be unloaded after completing a firing round before the shooter turns from the firing position or sets the gun down.

Accidents with guns may be fatal, and accidents don't happen, they are caused.

Everyone should know and practice firearms safety rules. It is absolutely necessary Security Personnel know and practice firearms safety rules.

If Security Personnel take guns home, or carry them in a family car, the spouse and children must know and practice safety rules. A gun is a deadly weapon and is to be treated as such. *A Gun Is Not A Toy!*

Restrictions on the use of guns by Security Personnel are many and should be known by all who bear arms. The National Rifle Association has many publications regarding firearm safety and they are available upon request. A short statement of these rules would include:

When on duty, Security Personnel may fire guns:
1. On the firing range.
2. To protect the life of self.
3. To protect the life of another.

FIREARM COMMANDMENTS

All Security Personnel should burn the following commandments into their minds — they must never be forgotten:

Commandment Number 1 —
"NEVER POINT A GUN AT ANYONE UNLESS YOU ARE PREPARED TO TAKE A LIFE."

Commandment Number 2 —
"YOUR GUN IS YOUR RESPONSIBILITY AT ALL TIMES WHETHER IT IS IN YOUR POSSESSION OR NOT."

CHAPTER IX

NATURAL HAZARDS

NATURAL HAZARDS TO SECURITY

No one is able to control the natural hazards to security. The only thing to do is to be prepared for such hazards and act quickly when such hazards occur. There are a few exceptions, but most hazards to security from natural causes are forecast by conditions. However, the warning may be of short duration, or come just before the hazard, so actions to limit the resulting damages must be pre-planned.

SEVERE COLD

The damage from severely cold weather can be extensive but precautions against such damage are possible. Security Personnel should have complete knowledge of heating systems, location of fuel supplies, location of cut off valves for water, gas and electrical panels. The location of tools to be used to help employees and others trapped by snow, ice, and other cold hazards must be known and easy to find.

Jumper cables to start automobiles, snow shovels, ice scrapers, extra tire chains and other emergency equipment should be in good condition and convenient for ready use. First aid is often needed for frost bite, heart attacks caused by snow shoveling, etc.

SEVERE HEAT

Heat damage is more of a natural hazard to human bodies than it is to property. However, during extremely hot weather there are potential damages from spontaneous combustion fires. Danger also exists in relation to other fires and from breakdowns of air cooling systems. In addition to these hazards, human conduct is often crazed during heat spells and much damage to property as well as other humans results. The ease of combustion during hot spells must be considered and fire fighting in hot weather is a difficult task. First aid, which is discussed in detail later in this book, is often needed during hot weather. Employees, customers, visitors and others are often overcome by heat exhaustion, apoplexy and other physical problems triggered by soaring temperatures.

HEAVY RAINS

Heavy and extended rains, especially those with high winds, can cause very heavy damage. The leaking roof with its dangers to files, machines, furniture and equipment is a major hazard to Security Personnel. It distracts Security Personnel from other duties and becomes a cover for law violators. The erosion of parking lots, the damage to partial and temporary structures, and standing water are hazards Security

Personnel must battle. The break down of communications during heavy rains is common (telephone — teletype — radio — alarm systems — CCTV) and gives persons intent on committing criminal acts a curtain to hide behind.

WINDSTORMS, TORNADOES AND ELECTRICAL STORMS

The natural hazards to security in the following paragraphs are lumped together since they often occur during the same time period.

Severe windstorms tax security forces since it is a hazard tacked on to other security hazards. Windstorms blow out glass, rip away marquees, down power lines and communications lines and often cause electrical fires. The shut down of power to elevators, falling trees, flying debris and dead traffic signal lights add to the confusion.

The second and more severe form of windstorm is the *tornado*. Tornadoes do tremendous amounts of damage to properties and are a threat to human life. The tornado watch warning, if and when received, should be heeded without fail. A plan of action to be followed in case of tornadoes should be on hand and constantly updated. The aftermath of a tornado is not a pleasant sight and rescue work calls for steady hands and cool heads.

Electrical storms and lightning are deadly hazards to security. The damage to electrically driven equipment is great. Security Personnel are under greater strain than usual following electrical failures. Many routine activities of the guarded business change from machines to human hands. The change results in slowdowns and errors. The loss of power and damages from surges of power invites those who would breach security. Starting emergency procedures is a must.

Fires from lightning strikes are additional hazards to the security system and emergency plans to combat them must be followed.

HAIL

Hail is most harmful to agricultural interests. The loss to crops is huge and there are other security problems resulting from a hailstorm. Broken windshields, windows, skylights, awnings, etc. offer opportunity to the criminal mind. Anything that takes the security force away from its regular patrol leaves a chink in the armor of protection.

FLOODS

Rising waters, with the evacuation of people, property being moved, and records being transferred help to breach security. The security staff must have, and must follow, plans designed to cope with such a disaster. Looters prey on flood areas and security is hard pressed to fight off these scavengers. When a major flood occurs, the National Guard is called out to help victims and to crack down on looters. *Help* is often a longtime in arriving, and until it comes security forces must take charge.

EARTHQUAKES

Security Personnel have their hands full following an earthquake. Foundations have been weakened, walls fall, live electrical wires and broken water mains add to the general problem. The most danger to Security Personnel comes from criminals who use the disaster as a cover to loot and rob. There is no prevention possible. The damage and danger that follows an earthquake can be lessened by planning and following the plan when the event takes place.

HURRICANES

All coastal areas of the U.S. can expect hurricanes. The states in the southern half expect the most. All states bordering the Gulf of Mexico are subject to hurricanes during the summer.

With modern electronic devices and aircraft watches available, advance warning about hurricanes is available. Hurricane advisories are issued at frequent intervals. With modern tracking systems, there is no reason to have people underfoot during a storm. Security Personnel should see to it that a plan for evacuating people before, during, and following a hurricane is on hand.

Hurricanes leave electrical wires hanging, communications lines down, areas flooded , trees and shrubs scattered, broken windows and disabled cars on streets. In many cases, severely damaged buildings with broken foundations clutter the landscape.

To Security Personnnel hurricanes mean looters, robbers, and others out to take property belonging to someone else. The hours put in during these times are difficult and nerve wracking. It takes all the ability the Security Person possesses to keep control.

As with other natural hazards to security, there is no way to prevent a hurricane — the only thing to do is to plan for its arrival and stay with the plan when the hurricane arrives.

NATURAL HAZARD SECURITY PROBLEMS

SEVERE COLD

1. People leaving early, all bundled up, may be stealing.
2. Physical danger exists from ice, sleet, snow, slippery walks, parking lots and roadways.

3. Telephone and electrical lines may be down.
4. Water pipes may freeze and burst.
5. Employee cars may not start.

SEVERE HEAT

1. Humans may suffer heat strokes and apoplexy or become exhausted.
2. Air cooling and conditioning systems need to be watched.
3. Fire is easier to start and more difficult to extinguish.

HEAVY RAINS

1. People wearing raincoats and using umbrellas may hide objects.
2. Parking lots and walkways and roadways may suffer erosion.
3. Cars slide off roads and become mired in mud.
4. Telephone and electrical lines go down.
5. Automobiles and other vehicles may drown out and need starting.
6. Streets and roads flood — water may get into buildings.

WINDSTORMS AND TORNADOES

1. People tend to panic.
2. Telephone and electrical lines go down.
3. Plans for evacuation must be followed.
4. Removal of debris becomes a problem.
5. Fire danger is always present.

ELECTRICAL STORMS

1. People are afraid of lightning — some may panic and need calming.
2. Telephone and electrical lines go down.

3. Fire is apt to occur due to striking lightning.
4. Computers suffer from overloads and power surges.
5. Air cooling, conditioning and heating systems need monitoring.
6. Unless back up power is available, loss of automatic doors, alarm systems and lights are probable.

HAIL

1. Glass damage, building damage, and windshield damage are probable.
2. Landscaping suffers.

FLOODS

1. Evacuation plans must be followed.
2. Looting is to be expected.
3. Parking areas will be useless.
4. Building foundation may be undermined.

EARTHQUAKES

1. Keeping posted will be difficult.
2. Ruptured utilities are probable.
3. People will panic.
4. Looting will be common.

SPECIAL HOT WEATHER HAZARDS

Hot weather brings with it special hazards to security. It is well known that humans commit more irrational acts during hot weather than at other times of the year. The heat affects Security Personnel and law abiding citizens as well as the criminal element. In view of this, the following items should be remembered for future use.

1. Heat Stroke results when body temperatures reach 106 degrees and remains there for any length of time.
2. Heat Cramps in the abdomen and legs may occur.
3. Heat Exhaustion brings on fainting, dizziness, weakness and a feeling of nausea.

Prevention of problems brought about by hot weather is better than attempting treatment after problems arise. A few recommended methods of preventing hot weather ills:

A. Drink plenty of water and/or other liquids. When thirst demands one glass of water, drink two!
B. Use enough salt to replace that lost to persperation.
C. Schedule the heavy work of the day before 10 a.m. and after 2 p.m.
D. Wear light colored and loose fitting clothing if possible during hot weather.
E. Do not drink alcoholic beverages.
F. If heat stroke, heat cramps or signs of heat exhaustion appear, sponge body surfaces with cool water.
G. Avoid hot places if possible — especially indoors.
H. Remember — tempers flare during hot weather — and that includes yours as well as others — be careful!

CHAPTER X

OUTSIDE AREAS

PHYSICAL SECURITY AREAS

There are a number of physical security areas Security Personnel must be familiar with, in detail. Included are:

Gates — Doors — Communication Centers — Control Desks — Consoles — Guard Houses — Entrances — Exits

POINTS TO CHECK

Damaged fences
Materials or articles stacked against fences/walls
Defective or open locks
Evidence of breaking and entering
Plant growth obstructing view and furnishing hiding places
Soil washed from under fences and walls

The boundaries, or perimeters, assigned to be guarded, must become as well known as the backyard at home. This is for the safety of Security Personnel as well as for the protection of the employer.

Chain link fences and gates, with any added features designed to protect the property, need to be checked on the first patrol of the shift. An open gate, a broken lock or other damage to the fence should be reported as soon as found.

Moats, if any, should be checked on the first round of the shift. The use of boats or rafts and fallen trees afford the criminal an opportunity to cross into sensitive areas.

Solid walls make good barriers but they also give the criminal cover from which to work. Examination of the top of the walls and entrance or exit breaks in the walls should be made early in the shift. Solid walls are scaled by using ropes and climbing up over items piled against outside walls. The use of fork lifts, front end loaders, cranes, and derricks on movable vehicles to lift thieves over walls is a common practice.

Gates are vulnerable to breach due to their nature. Constant surveys of all gates is mandatory.

Turnstiles must be manned at all times during the shift. Vehicular gates may be locked or otherwise blocked but they must be kept under watchful eyes at all times. If vehicles are entering or leaving during the shift, Security Personnel should be alerted so persons bent on crime or mischief do not enter.

The area security plan must follow a definite set of rules and regulations. If they are to be changed in any way, Security Personnel must be notified of the changes.

Outside storage areas must be checked on a constant basis so items are not left in, or near them, to be stolen at a later time. The area must be constantly checked to make sure it is being used properly.

Truck parking areas must be given special attention. Trucks moving in should be examined for contents and while on the yard kept sealed until loaded out under the watchful eyes of Security Personnel. Using trucks to move stolen items off the premises is a favorite criminal operation. Security Personnel are often the first to learn of such illegal activities.

EMPLOYEE PARKING AREAS

Employee parking areas should be kept separate from other parking facilities. Constant checks should be made for vehicle passes and/or decals. The area should be enclosed to help prevent robbery, muggings, and vandalism. There should be a central exit so unannounced inspections may be made.

The employee parking area should be protected by fences if possible, and kept for employees only. As stated previously, employers should not let employees visit their cars during working hours. It offers an opportunity for the crooked employee to move goods from the work area for later transport form the premises. The area should have special entry and exit points to be use by all personnel. If policy states all vehicles are subject to spot check, a central exit point gives Security Personnel the opportunity to make these checks. Once spot checking is in effect, losses from the parking area will almost disappear. No one wants to take the chance of being caught in public, moving stolen goods.

EMPLOYEE RECREATIONAL FACILITIES

Many businesses have employee recreational facilities. The areas used for these facilities are points dishonest personnel use to hide away stolen goods. Loosely fitting sweat shirts, baggy sweat pants, gym and golf bags, ice chest and other items normally used for recreational purposes become containers to carry out stolen goods. In addition, goods hidden during working hours in the recreational areas may be removed with ease when the culprit returns under the pretense of play. In addition, criminals use athletic clothing as a disguise to move about without identification.

If recreational facilities are within protected areas and if access to and from the area is to be permitted from the protected areas, the recreational area should be separately fenced and controlled by key cards or Security Personnel. The best situation is for the recreational area access to be from other than the protected area. The family, friends, and others who have access to the recreational area are threats to security.

BUILDING WALLS

Building walls are a part of the protective process. However, walls have openings, and these openings may be used by criminals for entry. Any opening large enough to allow human entry must be secured. Entry through them must be controlled during business hours. Fire exit doors with panic bars to allow emergency exits must have some type of alarm system on them to insure they are not used as entry points. Windows should have some type of barriers installed. Bars, heavy screens and alarm systems are desirable. Security Personnel making rounds will report damages to any of the protection devices and any apparent attempts that have been made to gain entry.

Any opening in the outside walls of buildings within the protected area should be treated as is the perimeter gate. ANY AREA THAT HAS MORE THAN 96 SQUARE FEET OF DIAMETER, THAT IS LESS THAN 18 FEET ABOVE GROUND AND LESS THAN 14 FROM ANOTHER STRUCTURE, SHOULD BE PROTECTED.

Fire rules should be posted on doors, alarms, etc. Windows should be secured with chain links, bars, or other materials.

Every building with a different use should be considered in the security sense in a different way. The building used to house steel I-Bars for construction will require different security measures than the building used to house computer terminals. The building in which the payroll is kept requires more security than the building housing the sales department. A survey, weighing the needs of each building in relation to total security should be made available to Security Personnel.

LIGHTING

"Let There Be Light" is a good motto for Security Personnel. Darkness is a cover for the predator, the arsonist and other criminals. The lighting of premises will not in itself provide security, but it will take away one of the favorite areas of the criminal — the area of darkness.

SECURITY PERSONNEL ON PATROL

1. Check lights over entrances — aisles — stairways — (Unscrewed or turned off lights between rounds indicates an intruder on the premises and a search is needed.)
2. Notify maintenance of burned out lights.
3. Know cover of darkness is the time for many criminal acts.
4. Know light on the subject is a must. (A lighting survey by a professional is a must for owners.)
5. Check manual off-on switches.
6. Change timing devices through out the year to keep up with changes in hours of daylight.

OUTSIDE STORAGE AREAS

Raw materials and partially constructed products are often stored outside. They must be properly stored and not put into a "junk pile." They should be covered for protection against the elements. A permanent storage space should have tie down facilities for canvas or plastic covers.

ROUNDS CLOCKS

Some patrols have "rounds clocks" for Security Personnel to "punch" at visits to the area of the clock. The purpose is to let the shift supervisor know the patrol is being made. In addition, a definite time factor is known when an investigation is made of an incident that happened during the shift.

The idea is a good one, but certain precautions are necessary in order to keep it from backfiring. One is — clocks should never be punched in a routine order. To do so gives a watching criminal the chance to time patrol rounds and to act accordingly. Staggering times to approach clock areas, having the clock punched several times in a row in an area, omitting an area on different patrols plus other irregular checking in at area stations will help keep the criminal off balance.

Keys to the rounds clock should be in areas open enough for Security Personnel to visit without giving protection to someone hiding nearby. Clock areas should be spaced so as to require complete coverage of the protected area. Locations of clocks should be changed from time to time so areas visited will vary.

Keys should be secured at area stations in a manner to prevent them from being removed.

Security Personnel should approach key areas with great care and thoroughly inspect the general area before approaching the key location.

SECURITY COMMUNICATIONS

Security Personnel must be provided with communication devices if top performance is expected. Instant response to alarms, accidents or incidents is the prime reason for having Security Personnel and each must have a way to communicate.

The best all-around communication device is the two-way walkie talkie with ability to connect wide ranging Security Personnel to the dispatcher and/or main office.

Telephone beepers are sometimes used when telephones are nearby. While they are good alternatives to walkie-talkies, they are not as good in overall effectiveness. A third method is to have roving Security Personnel call in from time to time. This method of communication is not good but is better than having the patrol out of touch altogether.

If the walkie-talkie system is used, a method of keeping batteries at peak performance at all times must be used. If each post on each shift is assigned a numbered and color coded walkie-talkie and it is placed in position to be re-charged after each shift the problem will be solved.

Whatever system is provided, it is most important the supervisor of the Security shift have two-way communications with the dispatcher at control center at all times.

VEHICLE CONTROL

All vehicles entering the Security area should have their movements carefully controlled. If the vehicles are trucks moving in and out of the area they should be clocked in, clocked out, and a watch made of their movements while in the area. This may be accomplished by use of Closed Circuit Television, by observation points in the area, or by use of a Security rider.

If the vehicles are rail cars, they should be brought into the area by railroad crews and sealed by the crews before they are left at docks or sidings. Security should check to see if cars are properly sealed and check undercarriage to keep stolen items from being hidden on the rail car.

At all loading docks, truck or rail, Security must keep a close eye on the areas under the vehicles for pilfered items.

Vehicle control should include the private vehicles of employees and management. It is highly important that each employee be advised of the exact rules, and penalties, for violations of rules in the parking and use of personal vehicles on company premises. It must be clear to them that Security is entrusted to enforce the rules, impartially but firmly, and policy of the company will dictate proper actions.

All restrictions on parking, speeding, restricted areas and the like should be made clear to all employees and notice given that Security will follow the book in case of violations.

CHAPTER XI

ALARMS

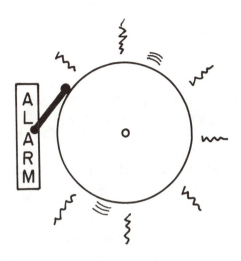

DETECTION AND WARNING SYSTEM

These systems assist security forces who know how they operate. They must be efficient. When used properly and installed properly, they are of great assistance in combatting the criminal. They are used in various ways and come in various forms.

FACTS ABOUT SYSTEMS

1. Fire prevention is made easier by the use of heat sensors and smoke alarms.
2. Breaking and entry robberies are often prevented by use of warning devices.
3. Detecting systems can trigger water to flow into extinguishing systems.

4. Water and gas leaks can be discovered with detectors.
5. Electrical, heating and air conditioning systems are monitored.
6. Foil circuits when broken sound alarms.
7. Invisible ray installations make evasion difficult.
8. Ultrasonic sound systems are good warning systems.
9. Microwaves are used in alarm systems.
10. Safe protection is assisted by use of electronic systems as are vaults.
11. A good system should be demanded by owners. Cheap systems are not assets.

In purchasing general detection and warning systems, the buyer should get a good system, test it, and know how it operates. It is wise to know its good points and its bad points.

ABOUT THE SYSTEMS

1. All openings in the perimeter to the protected area should have an alarm system to cover it.
2. All detection and alarm systems must be tested periodically.
3. Power sources should be hidden, protected, checked and tested continuously.
4. Employee designations should be made as to whom to notify at alarm.
5. Proper installation of detection and alarm systems bring lower insurance rates.
6. An alarm should be adequate for the business to be protected. If not, a false sense of security is established and the criminal moves in.
7. All employees should be aware of the operation of the system.

ELEMENTS TO BE CONSIDERED
IN A SYSTEM

1. Use of hardware — including electronics.
2. Utilization of manpower.
3. Use of software.

A study should be made as to whether to employ door and window intrusion detectors, whether to use photoelectric and laser alarms for perimeters, and whether to use vibration, motion detection or proximity devices, or perhaps use all of them.

Fire prevention systems (sprinklers) signal when heat hits certain heat points and trips the automatic flow of water.

Detectors are devices that are triggered when a break occurs and sounds an alarm. Some detectors transmit data when triggered. Alarm systems are needed to alert security forces.

SYSTEMS THAT MAY BE USED

A. Waterflow alarm system.
B. Surveillance of fire systems.
C. Heat detecting systems.
D. Heat detection systems tied to smoke detection systems.

INTRUSION AND BURGLARY SYSTEMS

1. Foil circuit protection system.
2. Invisible ray protection system.
3. Ultrasonic sound sensor system.
4. Microwave protection system.
5. Electronic safe protection system.
6. Vault protection system.

Systems have been developed to control building heating and cooling systems.

Pressure alarms, which give a signal when weight is brought against them, is a common but effective alarm. Used under windows, in front of art displays, inside doorways or as used in the gasoline station — stretched across the approach to the tanks — they give evidence barrier has been touched.

Another popular and effective alarm is the magnetic contact alarm. In this type alarm the signal is set off when a magnetic field between contact points is broken. A good example is the door with one contact point in the edge of the door proper and the other contact point in the door frame, when the door is opened the magnetic field is broken and the alarm sounds.

Alarm makers have become more expert in their field. Modern motion detection alarms have come into use. The two basic systems include radio frequencies or ultra-sonic sound waves to cover an area. When there is motion in the protected area the established patterns are upset and this causes an alarm signal.

A very sensitive alarm system is sometimes used in which a group of microphones are installed in an area. Sounds in the area are picked up and an alarm signal is given. The system is a good one, but the mikes are prone to pick up sounds outside of the area being protected. This may result in a false alarm. The sensitive alarm system is put to good use in vaults and other areas where sounds from the outside are not a problem.

The Photo-Electric Alarm system, which goes into operation when something passes through a beam between a sending and receiving cell, is in common usage. The system is ordinarily used to open and close doors in business places.

Strategically placed photo electric alarms become effective in signaling the presence of unauthorized persons.

Glass windows and doors are favorite entrance points of the criminal. Metal Foil has long been used in the alarm system. A break in installed foil sets off the alarm. This system is subject to problems brought about by age, and scratches in the foil, that result in false alarms.

Vibration Alarms are seldom used in the modern world. The system was based upon a set degree of vibration, which, when exceeded, set off a signal. Modern trucks, jet airlines and earth tremors usually exceed the level of vibration needed to trigger them.

Capacitance Alarms are commonly used when the object to be protected is made of metal. Safes, metal file cabinets and strong boxes are surrounded by a magnetic field a foot away from the outside surface. When the magnetic field is broken, an alarm is sounded.

FOUR ALARMS IN COMMON USE

Local Alarms — These alarms usually have a siren, bell or other loud noise built into the system. They are designed to be heard in the immediate area. These alarms are good if anyone, Security Personnel or employee, is near enough to hear it. It often goes off and no one hears it. In everyday practice these alarms are operated on batteries. If batteries are allowed to run down, or no one hears them, they aren't effective. (Security Personnel must know where alarms are installed and must make sure the batteries are always fresh.)

Proprietary Station Alarm Monitors — These are alarm monitors located within the main guard force office. When the alarm is activated, a signal is sent to the main panel. By sound or sight, sometimes both, the panel tells the monitors the location of the alarm area. This alarm does not sound at the point of the trouble, intruders may not know an alarm has been set off and the chance of catching them is increased.

Central Station Alarm Monitors — These alarm monitors are located in the office of the servicing company. When an alarm is set off, the signal is sent by telephone lines to a central location and appears on the main panel at the central station. The location of the property on which the alarm has sounded is noted and Security Personnel are sent to the location. At the same time, the Security Personnel on-site are notified and law enforcement agencies are notified. The weakness of the system is that time is lost from the time the alarm signal is received until Security Personnel appear at the alarm area. *However*, if the alarm received at the central is a fire alarm the fire department is notified and this is a great plus for the central station alarm monitoring system.

Remote Alarm Monitors — Remote alarm monitoring is the alarm signal sent directly to the local police or fire department. The signal is sent via telephone lines. This method of sending alarms is not available to most residential or commercial users in the private sector since police and fire departments do not have the manpower to monitor the alarms. These alarm monitors are used primarily for the protection of government-owned facilities.

Remember, an alarm system is only as good as the response to that alarm. The best monitoring system is one that uses on-hand Security Personnel — proprietary or contract — they can, and do, respond at once. Security Personnel entering the duty shift should make sure all alarm systems are on line, and if not, take steps to see they get on line.

If the alarm system being used depends solely upon electrical power, a back up system, powered by batteries, should be in place to automatically take over in case of power failure. In addition, a mechanical line supervisor should be installed to maintain a constant flow of low voltage current through the lines. if efforts to by-pass, or cut the wiring to the alarm system are made, the voltage level on the line is changed, and an alarm is sounded that indicates tampering or electrical system problems.

FIRE PROTECTION ALARMS

Security Personnel are the first line of defense against intruders and the first line of defense against the most destructive of hazards — *fire!* As a result, Security Personnel should see to it the fire alarm protection system has the three basic elements which give immediate detection and responses in fire situations.

1. Manual Fire Alarms — Commonly seen on buildings and along streets. Manual fire alarms should be well marked and easy to get to by all workers within the area to be protected.

2. Fire Detecting Devices — These devices come in a number of types. The Smoke Detecting Alarm operates much as the photo electric cell alarm. Smoke between the light souce and the receiver breaks the beam and sets off the alarm. These detectors are very effective in air conditioning and air vent ducts and are very popular in homes as fire alarms. *The batteries should be checked on a continuing basis.* The Ionization Detector operates when a small amount of current is passed in the air betweem two plates. A fire in the starting up stage gives off hydrocarbons which disturbs the passage of the current and sets off the alarm. The use of this type detector in areas where spontaneous combustion is a danger, such as computer centers and hospital areas, is a must. Due to the early warning it gives, it is a good choice for any fire protection system.

3. Heat Sensing Detectors — These detectors operate on changes in temperature that are rapid or rise

above certain levels. When the change in temperature occurs, the alarm is sounded.

Supervisory Alarms — These alarms support or accompany sprinkler systems. The six major types are:

a. The water flow alarm in the main riser has a paddle in the pipe leading to the sprinkler heads connected to a switch. When the sprinkler head is activated by a ruptured head in the wet system, the paddle moves, the switch is thrown, and an alarm is sounded.

b. The air pressure alarm sounds when the air pressure in a dry system or storage tanks drops below recommended levels.

c. The water level alarm sounds when the storage water level drops below required levels.

d. The temperature alarm sounds when water nears the freezing point.

e. The post indicator alarm sounds when the main water riser to a valve has been shut off.

f. The gate valve alarm sounds when the secondary water valve has been closed.

REMEMBER — All the fences, alarms, safeguards and other protection devices on premises will be worthless if Security Personnel are not properly trained to use them.

REMEMBER — Detection and warning systems are made to assist personnel, not to replace personnel and they make personnel more efficient.

CHAPTER XII

PHYSICAL CONTROLS

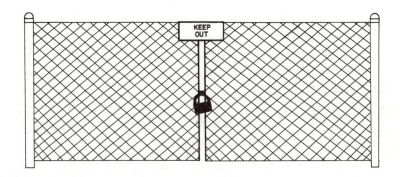

PERSONNEL I.D. AND CONTROLS

If physical security of an area is to be firmly established, Security Personnel must know:

1. Who may enter?
2. Where may they enter?
3. At what times may they enter?
4. Who has keys to let themselves in from unguarded entrances?
5. Are visitors required to have escorts?
6. Who signs I.D.'s and who can make exceptions?
7. Who can autorize company property removal?
8. Who is to be notified if company employees refuse to obey I.D. rules?
9. Who is responsible to tell employees of I. D. rules?
10. Will Security be backed up when reporting I.D. rule violations?

It is estimatied that one of every ten employees in the U.S. steals from their employer. This ten percent of the work force opposes the requirement for identification and movement control. They enlist the support of others in order to do away with such rules. These thieves prey on the minds of the others in the group. They make them feel they are not being trusted. There is another group that merely follows these thieves and their cohorts.

A GOOD PERSONNEL I.D. SYSTEM

1. An I.D. Card with photograph.
2. The name, social security number, date of birth, physical description and signature of the person issued the card.
3. A clear plastic laminated card to help prevent forgery.

Any time company property must be moved outside the security area, a property removal pass should be issued. The pass should contain all information regarding the property so Security can check it quickly. The pass should be made in duplicate so Security has a copy to attach to the daily report.

Visitors need to be identified upon admittance to the Security area. Some visitors should be escorted. Others should be allowed to visit the area to conduct business on their own. Whatever the rules, Security should be aware of requirements and follow them without fail.

THE VISITOR LOG SHEET

A. Visitor's name and company represented.
B. Time the visitor arrived and the time the visitor left.
C. The person or department visited and reason there for.

D. Any remarks as to the visit deemed important such as any equipment issued, items left at the entrance station, etc.

LOCK SECURITY

It has been said that locks are used to keep honest persons honest. Although this may be true, locks also act as deterrents to those who make attempts to get past them. The time a lock deters a thief may be the factor that protects a property from considerable damage or loss. A brief description of locks and comments on them:

1. Disc Tumbler Locks — Used for automobile doors and trunks, desk drawers and file cabinets. The time required to breach is about five minutes. They are not good safeguards.

2. Pin Tumbler Locks — Depending on the number of pins used and the materials used, these locks have poor to average ability to protect. They are generally used in residential and industrial areas.

3. Barrel Keyway or Circular Locks — This type of lock gives pretty good protection since it is hard to pick. It is a good lock to use in the overall security system.

4. Level Locks — Depending on the material used and quality of construction, this lock is used for many purposes. Safety deposit boxes use a heavy duty grade and cabinets and desks may use them.

5. Maximum Security Locks — A pin tumbler type of lock that is very hard to pick. It was developed to provide for pins to interlock in the keyway. Maximum Security is a trade name.

6. Combination Locks — These locks are difficult to open by other than experts and give good delay times. The combinations should have four or more numbers and should be able to be changed by other than a locksmith.

7. Cypher Locks — These locks use combinations of buttons or letters that must be pressed in order to be opened. These locks, like combination locks, offer good delay times in opening.

It should be noted that the door the lock is used in must be solid so as to discourage entry. Hardware must be of first quality and properly installed. Dead Bolt locks in door jambs are excellent guards but they must be of good quality and properly installed.

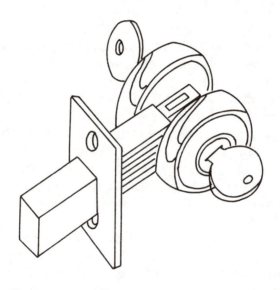

PADLOCKS

These locks, combination or keyed, are used in many cases. Their effectiveness in delaying intruders will depend upon:

1. The shackle. Hardened steel is the only satis-factory material.

2. The material and construction of the lock.
3. The material and construction of the hasp and how the hasp is installed.

Key locking devices must be backed up by other security measures. The glaring weakness in the system of key locks is in the key itself. A thief with a key can open a locked device, close it, and unless otherwise detected, be on the premises without knowledge of Security. This is why *Key Control* is a must. Positive control of keys must be maintained.

KEY CONTROL CHECK LIST

1. Key cabinet and key depository is part of the security system.
2. Key record must be kept — only officials to have them.
3. Key blanks with code numbers must be recorded.
4. A constant inventory must be kept of keys, key blanks and duplications.
5. Frequent audits must be made of key control records.
6. A daily report of who has keys is mandatory.
7. Master Keys must be accounted for at all times.
8. Tags for keys are a must. They can be color or otherwise coded.
9. Action must be taken upon report of a lost, damaged or stolen key.
10. A daily issue and receipt for keys is priority.
11. Locks must be rotated - at least annually.
12. Combination locks must be changed very often.
13. Remember — key locks are easily picked.
14. The conventional combination lock is of little value.
15. Re-locking devices must be checked often.
16. Locks with interchangeable cores are good for security.

17. Push-button combination door locks are excellent.
18. Locks bolts must be flush pointed inward.
19. High grade steel hasps are desirable.
20. Keys should have the words (Do Not Duplicate) on them.
21. Restrict Master Keys to Security Forces.

CLOSED CIRCUIT TELEVISION (CCTV)

A CCTV system combined with a microphone and speaker at selected camera locations is most valuable.
Benefits of CCTV:
1. Cuts down accidents due to carelessness, recklessness, horseplay.
2. Production efficiency increases.
3. Keeps the temptation to pilfer down.

COMPUTER SECURITY

Inside dangers to computers:
A. Theft (Tapes — Time — Etc.)
B. Espionage (Stolen Data)
Outside dangers to computers:
1. Fire
2. Sabotage
3. Industrial Accident
4. Mechanical/Electrical Malfunction
5. Natural Disaster
Computer users cannot depend on substitutions. The reasons are many and include:
1. Incompatible machine configuration.
2. No open time available.
3. Software compatibility limits.
4. Information flow may not mesh.
5. The alternate may also be down.

VISITORS

Visitors may be friends, family, salespersons, or information seekers. Regardless of classification, all visitors should be viewed within the boundaries of company policy.

1. Who may visit?
2. Where may visitors be allowed to go?
3. Who issues visitors passes?
4. Are color coded passes available?
5. Should visitors have I.D. Cards?
6. Is the person and the visit a valid trip?
7. What time are visitors allowed?

There are many kinds of *visitors* and the alert Security Person should be aware of them.

CHAPTER XIII

SPECIAL SITUATIONS

RIOTS AND CIVIL DISTURBANCES

Riots are considered to be the wild and almost uncontrollable disturbances created by large groups of people. The mental image is that of rock and brick throwing mobs, assaulting persons and property, burning, looting and refusing to obey authority.

Strikes are considered to be the acts of members of a group, usually unionized, to refuse to work for employers until certain demands are met by management. Most strikes are non-violent picketing of premises of the employer. However, some strikes have blossomed into full-blown clashes with rioting a common occurrence.

Sit-Ins are generally non-violent methods used by groups to make a point against authority. They consist of persons in the group taking a place on the premises of the authority. They remain there until removed, until the point is made, and grievances are settled. Sit-Ins sometimes deteriorate into near riots or riots.

Protest groups are composed of people with a single aim in a special field. Their acts are mainly the carrying of signs in parades and/or around properties owned by those with different views. Protest groups may become violent and cause riots.

Each of these forms of protest have been used for centuries. Peasant uprisings against kings and land owners were common. Slaves have rebelled against their masters ever since one person placed another person in bondage. Religion has been the basis for civil disturbances in many instances. The crusades of the middle ages a good example.

In recent times, in the U.S., civil troubles have been seen in the race riots in Detroit in the 1940's, the Attica prison riots in New York, the racial riots of Watts in Los Angeles, California and the riots in Florida, namely Liberty City and Overtown. The burning of college buildings during the 1960's and the throwing of rocks at President Nixon while visiting foreign soil are other examples.

Broadly classified, riots and other civil disturbances will fall into one of the following groups:
A. Social Protests — These may be prompted by racial problems, religious fervor or uprisings at sports events.
B. Economic Protests — These may be prompted by demands for higher wages with resulting strikes; or they may come from people living in poverty.
C. Political Protests — These incidents, in the U.S., are usually confined to parades with signs showing unhappiness, or the peaceful picketing of governmental premises. In other countries, these protests are marked by rock throwing, gunfighting, and sometimes the overthrow of the government in power.

D. Disaster Disturbances — Protests in times of disaster are usually in the form of demands for food, shelter and clothing. Looters use disasters as a cover for their thievery.

E. The Failure of Authorities to Maintain Control — When authorities fail to make attempts to keep the peace, there are those who take advantage of such failures. When authorities do not keep a firm hand on events, they quickly get out of hand.

Security Personnel are often employed during riots and other civil disturbances. They must act with caution, but with authority, when called upon in such situations.

Two definitions Security Personnel should burn into their minds are:

Crowd — A law abiding gathering in which individuals act and think like law-abiding citizens. (A 4th of July picnic crowd, for example.)

Mob — A gathering that takes the law into its hands, with individuals losing their sense of reason and blindly following some self-appointed leader. (A lynching party.)

Security Personnel, when confronted with a situation that appears to be turning into a riot, should call for assistance from law enforcement officials. Mob actions can be nasty.

MOB ACTIONS

1. Use of words to abuse and anger authority.
2. Attacks upon persons and vehicles.
3. Throwing objects at persons and property.
4. Breaking glass, turning on water faucets and draining fuel supplies.
5. Starting fires to create fear and to block roads.

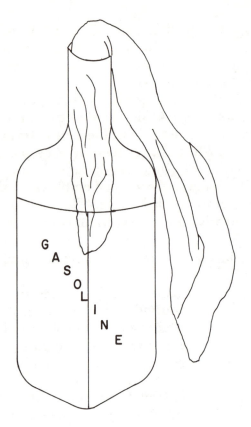

MOLOTOV COCKTAIL

6. Using dynamite and other explosive materials to create panic.
7. Bringing firearms into play and sniping.
8. Looting on a wide scale.
9. Using children and older persons as pawns against authority.

Security Personnel assigned to duty when riots or other civil disturbances are expected should report to a supervisor, usually a full fledged law enforcement officer. Orders given should be followed with attention to detail and to personal safety. *Do Not Become A Dead Hero!*

146

SECURITY AND GUARD DOGS

The use of guard dogs by Security Personnel is recommended in many situations. A good guard dog has keen ears, a sharp nose, and ears that hear more sounds than those heard by the best of Security Personnel.

German Shepherd or Belgian Police dogs are the best dogs to be used. They obey commands, like the work, and are the dogs most feared by the criminal element.

Guard dogs can be used to control humans because mankind still fears animal power. It is difficult to defend against the trained guard dog, especially when engaged in criminal activity.

Guard dogs used by Security Personnel can work on a leash if desired, or can be used as scouts in advance of patrol. In bad weather the guard dog can be used to great advantage since the dog is used to roaming and hunting under poor conditions. If noise level on the patrol is high, the work of the dog may not be at a high level.

The use of guard dogs in security work is not widespread due to turnover of Security Personnel.

Guard dogs can be trained to attack on command and can be trained to bite. They are very useful on patrol around junkyards, airports, lumber yards and in areas where footing is risky for the human. Competent guard dog handlers are hard to find and guard dogs must be carefully managed to prevent injury to innocent persons.

The use of guard dogs in warehouses, in between double fenced areas, and on cables between buildings is effective in preventing intruders from entering the premises. Guard dogs should be moved to different sites from time to time. Moving is easy because dogs can be transported in automobiles or light trucks.

A guard dog on a chain makes a good alarm. A free ranging guard dog on patrol can find openings in the perimeters and follow intruders to the area where criminal activity is taking place.

A guard dog is seldom named "man's best friend," but he or she can be a vital part of the security patrol. It is well for Security Personnel to remember it is safer to sic a dog on an intruder than it is to shoot a person.

SECURITY FOR SHOWS
(INSIDE AND OUTSIDE)

Security Personnel are often the only uniformed people working trade shows, boat shows, and fashion shows. They are also often the only people working concerts and other indoor entertainment.

Many show promoters have a Chief of Security. He or she ordinarily acts as the coordinator of security activities. They hire local Security Personnel, usually from an agency, to help during the set up, to work floor patrol, handle crowd control, direct traffic and parking, and to be on hand at the tear down and departure.

Security Personnel assigned to shows should "case out" the scene of the coming event. City auditorium, school gymnasium or whatever facility to be used should be physically checked out for Security control points.

Using a floor plan, all entrances and exits must be examined for locks, ramps, and crash bars. Those to be used by exhibitors, or other persons connected with the show should be noted on the floor plan. The method of identifying people entering and leaving the area should be determined.

The floor plan is valuable in spotting the location of all fire extinguishers (these should be checked to see if they are in working condition) and the type of fire they may be used on. An emergency plan to be followed in the event of a disaster should be a part of the security plan. Lanes established outside the building to allow emergency vehicles to enter are a must. The evacuation of people from the building, using whatever exits might exist after disaster strikes, should be known to all Security Personnel.

The areas with emergency lighting facilities, the placement of plants and decorative items, the location of concession and ticket booths should be observed and noted. Offices, conference rooms, utility areas and the rest rooms should be plotted on a chart.

The location of cut off valves for water and fuel lines (and how to operate them) is a must. The location of the panel box with circuit breakers for electric lines must be known. Repeat — *Know How To Cut Off Water And Fuel Lines And Know How To Cut Off Electrical Power.*

The Chief of Security should estimate the number of guards to be used during the show, and their classifications.

Security Personnel must know first aid areas, and the location and number of emergency medical personnel available. They must know the location of, and routes to the nearest hospital facilities. The location of first aid supplies and how to use them is a responsibility of Security Personnel.

The complete schedule of events must be studied. Security Personnel must know the time schedule for each event, the name of the person assigned each event and steps to be taken in case the shift relief does not appear.

Sometimes an event will be held outside. For example, turf grass show and equipment demonstration. Security Personnel, should "case out" the site of the event. Entrance and exits for exhibitors should be noted and the I.D. system to be used made a part of post orders.

The location of electrical lines and cutoff switches, the location of valves for control of water and other utilities and the precautions to be taken after dark are items to be considered.

If a tent is used, the condition of poles and stakes must be checked. The fire and wind resistance of tent materials used must be known. A plan for evacuation must be made in case of disaster.

In any type of show situation, inside or outside, Security Personnel must prepare for the worst, and hope for the best. All assigned personnel must be ready, willing and able to move with speed to prevent panic if disaster strikes. A spectator or exhibitor may have a heart attack or suffer other injuries. Handling these situations while the show goes on, is a result of advanced planning and carrying out plans. The responsibility of Security Personnel is to get the job done.

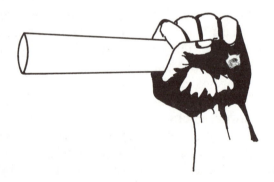

PARADES

Everyone loves a parade. However, staging a parade requires special security work. All Security Personnel assigned to parade duty should go over the entire parade route several days in advance of the formalities.

Entry routes into the staging area must be checked, the assignment of parade units and the order in which they leave the staging area must be known. The parade route should be walked with careful attention paid to details. The location of points available for crippled units to exit the parade, spots where difficulties might occur, such as intersections, bridges and sharp turns, must be investigated. Sites where emergency vehicles can be quickly reached must be platted. The location of emergency first aid stations must be memorized. The time it takes to get medical care to a stricken person is often the key to recovery.

Security Personnel assigned to parade duty must remember what to do in case of a riot. Almost all of the human hazards to security exist at a parade. Drunks, dopers and mental incompetents attend parades. Handling them quickly and efficiently is demanded.

Natural hazards to security may be present at parade events. The prevention of panic and the orderly evacuation of the parade route is the goal of Security Personnel if a natural disaster occurs.

Whether it is a show in a dinner theater, and outdoor exhibition of antique automobiles, or a parade, there is the possibility of poor human behavior. The problem of traffic control, parking, and disabled automobiles is ever present. It is because skillful Security Personnel handle these events that they run smoothly.

SPORTS AND OTHER STADIUM EVENTS

Lumping sports events, concerts, horse shows and other stadium events into a single category is done because of the similarity of situations faced by assigned Security Personnel.

Each event draws large crowds with many traffic and parking problems. Each attracts different types of people. The potential for violent human behavior is present at each event.

151

In addition, in each type of event hazards to security from natural causes exist. The hazards are somewhat reduced in the domed arena.

Football, soccer, baseball, basketball, ice hockey and other team sports have fans who seem to hang on the edge of fanaticism. They root for their teams with passion and sometimes become unruly. The use of alcohol and drugs at these events makes the job of Security Personnel more difficult.

All Security Personnel assigned to stadium or arena duty should review methods of crowd control and be prepared for anything.

ESCORTING V.I.P.'S

The escort of well-known persons is ordinarily carried out by Security Personnel in the employ of the V.I.P. The cooperation of local law enforcement officers in the area to be visited is always sought.

Presidents, Kings and other important governmental figures have the services of their secret service personnel available. Local law enforcement personnel cooperate with the secret service personnel to handle most details.

Security Personnel assigned to duties during V.I.P. visits are usually hired to work traffic, control crowds and handle parking. They are also hired to cover emergency situations. The well-*trained* Security Person who knows how to handle parades is a valuable addition to the security force when V.I.P.'s are in the area.

ELEVATOR EMERGENCIES

It doesn't happen very often, but when it does happen, Security Personnel should know what to do when an elevator stops and traps the passengers. It is a minor and unusual situation but it can cause a great amount of anxiety.

The first rule is — as it is with other emergencies — *Don't Panic!* Security Personnel must remain calm and help others to remain calm.

When a call comes from the stranded elevator via the emergency telephone, the switchboard operator should look at the outline of emergency procedures, furnished by the owner or Security Personnel. On this list will be the steps to follow for the handling of elevator emergencies. Suggested steps should include:

1. The operator should get all information possible from the caller. The approximate location of the elevator in the shaft, the number of the elevator, the number of people aboard and the physical and mental condition of the passengers.
2. The operator should allow the calling person to end the conversation and to hang up first.
3. The operator should tell the caller help is on the way and there is no need to worry.
4. The operator should immediately notify Security Personnel and maintenance personnel of the situation.
5. Security Personnel should go to the floor where the elevator is stuck and give assistance to maintenance.
6. When the passengers have been evacuated, Security Personnel should follow first aid recommendations if needed.
7. When the elevator has been returned to working order, Security Personnel should make a determined effort to learn the cause of the breakdown. Criminals may have been practicing tactics!

ARGUMENTS

There is only one rule to follow when arguments or quarrels start in a place of business. Security Personnel are empowered to, and should stop, any argument or quarrel as soon as it comes about.

If the argument or quarrel is between employees —
stop them immediately. If the argument or quarrel is
between customers — stop them immediately. If the
argument or quarrel is between employee and customer — stop them immediately. If the argument or
quarrel is between visitors and any of the above — stop
them immediately. *Arguments and quarrels lead to
major problems.*

FIGHTS

If a fight breaks out — stop it as soon as possible. Be
very careful when intervening in a fight — don't get
caught in the middle. If the fight is serious, call the
police. Make notes for the incident report with names
and addresses of all involved.

SAFETY

Security Personnel must make sure they have a safe
place in which to work. The saying, "Familiarity
Breeds Contempt" applies to the work place as well as
to people. Falls and improper use of equipment are two
reasons Security Personnel suffer so many injuries.

Slippery floors, falling over misplaced articles and
furniture, bumping heads on low hanging pipes and
falling down stairs are among the "causes" of so-called
"accidents."

Standing on chairs instead of using a ladder, opening gates and doors without thinking, improperly
handling fire extinguishers during examinations and
other supposedly "safe" tasks take their toll. *Think!*

ACCIDENTS

Security personnel must believe accidents do not
happen — they are caused. Security Personnel must
report, as they see them, conditions that cause
accidents.

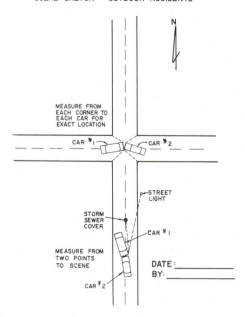

When an accident happens, first aid is given if possible. Calls for assistance should be made immediately. Ambulances — policemen — firemen — doctors — nurses — management and others named on local orders must be called.

Specific Information Security Personnel Should Have Before Accepting An Assignment

1. The location and types of fire extinguishers on the premises.
2. The location of electrical panels and cut off switches.
3. The location of water mains and cut off valves.
4. The location of fuel lines and cut off valves.

BASIC KNOWLEDGE OF FIRE FIGHTING IS A MUST FOR SECURITY PERSONNEL — FOR THEIR PROTECTION AS WELL AS PROTECTION FOR OTHERS.

Security Personnel Have A Duty
To Report, Without Prejudice

1. Reckless operation of equipment. Speeding fork-lifts, for example.
2. Failure to wear protective gear. Goggles when grinding, for example.
3. Use of fire or smoking in sensitive areas. Fuel storage for example.
4. Poor housekeeping practices. Mop buckets left inside doors for example.
5. Observed hazards. Overloaded electrical outlets, slippery walks, or shrubbery blocking an intersection are examples.
6. Employee conduct. Security Personnel must be diplomatic here, but incidents such as drinking, using drugs, horsing around or sleeping on the job are reasons accidents are caused.

Security Personnel are often exposed to situations that lead to accidents. On the first patrol round each shift it is wise to move slowly, as if it were the first round of a new assignment. Manholes left uncovered, ladders left standing, new ditches dug for utilities, and interior or exterior building changes may have taken place since the last round was made. *Be Careful — Accidents Are Caused!*

PUBLIC RELATIONS

Security Personnel work in a "people business." The number of people Security Personnel come in contact with on a "one-on-one" basis is limited. But the contacts that are made usually leave a lasting impression. The image the general public has of Security Personnel has been touched on previously, and steps to improve that image were discussed. By improving their image, Security Personnel perform one of the tasks in a good public relations program.

The relationship between the citizen and those who wear uniforms, or those who have more authority than others, is never an easy one. A certain amount of jealousy always exists towards the seat of power.

Security Personnel must constantly sell themselves, sell their company or employer and sell their profession. Their personal appearance, the appearance of the area they control,and their attitudes are always on display. They are always being judged by the public.

Security Personnel must be courteous to everyone. The rude word, the negative act, and the poor attitude is remembered far longer than a warm smile and kind word, the positive attitude, and the positive actions.

By being fair minded, and treating the public and fellow guards in an honest and friendly manner, a positive image of Security Personnel is created.

Every coin has two sides, and every situation has two sides. Security Personnel who make efforts to see both sides of a question make the wisest decisions.

Security Personnel must make a habit of showing themselves as able professionals. They must learn to be patient and friendly, whatever the situation. They must be the calm and brave figure of authority in the most trying of circumstances. The description, borrowed from fiction, "A fist of mail and a heart of gold," is one Security Personnel should try to copy.

A few minutes spent giving directions to a stranger, showing interest in the welfare of others and giving personal time to such things as little league, civic organizations and neighborhood activities is practicing good public relations.

Security Personnel represent themselves, their profession and all other Security Personnel, wherever they go. The better they handle the public relations portion of their job, the better off everyone will be.

Since the public relations aspect of the jobs undertaken by Security Personnel is so important, a few things bear repeating. These things should be committed to memory so they become a permanent part of the personality makeup of every practicing Security Person.

1. Always appear and act like a professional.
2. Be courteous in contacts with everyone.
3. Cooperate fully with fellow guards and law enforcement personnel
4. Work with those who are known as "opinion molders." They are many in number, and they have clout. The members of the Media (newspapers, radio and television), the Clergy, Teachers and Public Figures fall into this category.
5. Work as much as possible with groups such as Veterans, Minorities, Labor, Business, Management and Civic and Fraternal organizations.
6. Work with Government Agencies and Government Personnel but be very careful when dealing with them. Whatever level they fill, local, state or federal, they become bureaucrats — and never forget — The Bureaucracy Is Powerful!

CHAPTER XIV

FIRST AID

FIRST AID

All Security Personnel should have completed a basic first-aid course. If they haven't, they should ask their employer to contact the American Red Cross and arrange for such training. The Red Cross usually furnishes instructors and arranges the time and place. It would be wise to ask for CPR training at the same time.

The following information on first aid is given because Security Personnel are usually first on the emergency scene. They may also be the only authority present when a person needs medical help. The following information in no way takes the place of proper first aid instruction. It is for reference and to remind Security Personnel of events they may meet and must respond to as figures of authority.

The first rule for Security Personnel in first aid situations is to remain calm and take charge until medical help arrives. Preventing panic and keeping order may prevent the injured from going into shock. Most people, in emergency situations, respond to those who take command and will help bring order to the scene.

Security Personnel should remember their first act is to call for help then to go to the aid of the injured. The First Aid Kit should be taken to the scene. The kit should contain, as a minimum:

Two   2-inch compress bandages
Two   4-inch compress bandages
One   eye dressing package
One   ammonia inhalant (smelling salts)
One   package of 3 x 3 plain gauze pads
One   airway tube to use in mouth to mouth resuscitation
One   roll adhesive tape
One   pair of sharp scissors
One   supply of bandaids
One   bottle of antiseptic
One   pair of tweezers

Note: Security Personnel should check the first aid kit each time they go on duty. If the kit has been used a list of used items must be reported and replacements made.

Security Personnel are not trained medically and should never try to give any treatment unless absolutely necessary. They should give emergency treatment only if they know proper first-aid techniques.

There are five basic life or death emergency situations to be recognized and acted upon. They are all very serious and the sooner medical help arrives the better the chances are for the victim to live.

The first life-or-death situation is:

STOPPAGE OF BREATHING

If a person stops breathing, it must be restored in 3 to 6 minutes or the person may die. Whatever caused the breathing to stop, drowning, electrical shock, over dosing on drugs, choking or "swallowing the tongue", first aid action must be swift. First, get the head of the victim back and the jawbone moved forward. If breathing doesn't start at once, begin mouth to mouth artificial respiration. All Security Personnel should know how to do this. Quickly check to see if the victim has anything in the mouth or throat such as gum, food, dirt or other matter. Pull on the tongue if it has been swallowed, get it out of the throat. Use the Heimlich Maneuver to force food out of the throat.

The second life-or-death situation is:

STOPPAGE OF THE HEART

The treatment to restore flow of blood from the heart by external heart massage is known as CPR — Cardiopulmonary Resuscitation. Only a trained person should attempt this technique. Unless the CPR technique has been learned well and properly executed, it could cause damage to the victim. CPR trained Security Personnel are in great demand and usually get higher wages for having had the training.

The third life or death situation is:

BLEEDING AND HEMORRHAGE

A person can bleed to death in less than two minutes! The loss of one of the six quarts of blood found in an average adult body may be fatal. The quickest way to stop external bleeding is by placing pressure directly on the wound. It is fast, direct, and often the best method. There are three types of external bleeding. They are, with first aid control measures:

1. **Venous Bleeding** — Blood coming from the veins is usually dark red and flows in steady steam. To control, apply direct pressure to the wound with a compress bandage or handkerchief. Even a bare hand pressing the wound may give the blood enough time to clot. The time to clot is usually three to six minutes.

2. **Capillary Bleeding** — Blood coming from the capillaries oozes form the wound. Direct pressure on the wound with a compress, handkerchief or bare hand should stop the bleeding since capillary blood clots rapidly.

3. **Arterial Bleeding** — Blood comes form the arteries. It is bright red and gushes or spurts form the wound. This is the most serious type of bleeding and must be stopped quickly. The heart is pumping blood directly into the arterial system and blood loss is rapid. Put a compress bandage, handkerchief or the hand directly on the wound and keep pressure on until help arrives.

NOTE:

1. **Tourniquets** — The use of a tourniquet is not recommended. However, if a tourniquet must be used, the time applied should be noted and given to the first medical person available.

2. **Internal Bleeding** cannot be treated by untrained persons. If suspected, keep victim quiet and treat for shock.

The fourth life-or-death situation is:

POISONING

There are three classes of poisonings.

1. **External poisoning.** Caused by chemicals, gasses, acids or burning metals that affect the skin and/or eyes. For poisons on the skin, the affected clothing should be removed, the skin flooded with water, washed with soap and water, then rinsed re-

peatedly. Call the poison center or a doctor as soon as possible. For poisons in the eyes, the eyes should be bathed with warm water from a container for at least 15 minutes. The container should be held about 4 inches from the eye. Call the poison center or a doctor as soon as possible.

2. **Internal (Swallowed) Poisoning.** — Any number of items if swallowed can be fatal.
   a.   Medicines — Overdoses or mistaken doses should not be treated by mouth without medical knowledge.
   b.   Chemicals — Household Products — If patient is conscious and can swallow, give milk or water to dilute. Call medical center to see if vomiting should be caused. (Lye, bleaches and ammonia cause burning if victim vomits).

3. **Internal (Inhaled) Poisoning.** — Get the victim to a source of fresh air. Ventilate the area and give mouth-to-mouth resuscitation if needed. Call poison center or doctor.

There are many sources of poisoning. Medicines, chemicals and household products plus garage and garden products and a wide variety of personal products are common poisoning sources. In addition, there are a great many poisonous plants in the world. Poison control centers should be called any time it is suspected plant material has been swallowed. Many persons are highly allergic to poisoning from contact with plants such as poison ivy, poison oak, poison sumac. If exposed to such contact, the skin area should be rinsed, washed with soap and water, rinsed again and medical advise followed.

The fifth life-or-death situation is:

SHOCK

Shock is almost always present when a person suffers serious illness or injury. Security Personnel should presume the victim is in some stage of shock and take steps to relieve the victim. First aid for shock is:

1. Keep victim lying down with feet raised above the head unless there is head or chest injury.
2. Loosen clothing but keep victim warm. Do not apply heat but maintain the heat of the body.
3. Treat injuries but do not let the victim see the injuries.
4. Tell the victim help is coming — that all will be OK.
5. Remember that shock is a killer — the cardinal rule is to keep body temperature from falling.

THE SYMPTOMS OF SHOCK

1. Pulse rapid but weak.
2. Skin cool to touch — forehead clammy.
3. Lips and fingernails pale or turning blue.
4. Victim is chilling.
5. Victim is nauseated.
6. Pupils dilate.
7. Breathing is irregular.
8. Victim is thirsty.

Security Personnel are often called upon to render first aid in a number of special situations. While waiting for medical aid to arrive, Security Personnel may have to give first aid as follows:

UNCONSCIOUSNESS — The victim may be unconscious. Red face, fixed eyes and paralysis may be symptoms. Breathing may be irregular or even stopped altogether. The cause of the unconscious body may not be known. It could be due to stroke, heart attack, electrical shock, diabetic coma, concussion, exposure to extreme heat or cold or any number of other reasons. If the reason is unknown, first aid is to loosen the clothing, keep the body warm, and treat for shock.

FRACTURES — If a fracture is known to exist or is suspected, do not move if it can be helped. Keep the victim warm and dry. Treat for shock. If it is a compound fracture and there is bleeding, control the bleeding. If the victim must be moved, apply a temporary splint. Temporary splints can be made from rolled newspapers, canes, magazines, wooden or plastic stakes or other light materials. Place the splint around the break and past the joint above and the joint below if possible. (In lower leg above knee and below ankle — lower arm above elbow and below wrist.)

BURNS — If chemical burns — flush affected area with as much fresh water as possible. The cleaner the area becomes the better chance of recovery.

If scalding from water or steam, or flame burns are present, keep victim as comfortable as possible and treat for shock. *Do Not Apply Ointments, Creams Or Salves.*

As is the case with other duty assignments, Security Personnel must use a great deal of common sense when dealing with first aid situations. Security Personnel must remain calm. This instills trust. Victims must be kept comfortable as possible and kept reassured. Treatment for shock is always correct. Action must be taken quickly and with decision. As soon as help arrives, try to determine the What, Where, Why, Who and When of the situation. If witnesses are available, get their names and addresses. Make a detailed report as soon as possible.

Security Personnel are not medical personnel. If they have received training in first aid and CPR, as they should have, they are better prepared to give first aid. There are ten "do nots" for all Security Personnel to remember in first aid situations:

10 DO NOTS

1. **DO NOT** give any form of medicine.
2. **DO NOT** try to remove objects from the eye.
3. **DO NOT** give liquids to unconscious persons.
4. **DO NOT** lift injured persons to sitting or standing positions.
5. **DO NOT** move persons with internal injuries or fractures.
6. **DO NOT** put ointments, creams or salves on burns.
7. **DO NOT** restrain epileptic seizures.
8. **DO NOT** let victims see injuries.
9. **DO NOT** discuss injuries with unauthorized persons.
10. **DO NOT** give up a victim as dead until so advised medically.

CHAPTER XV

FIRE

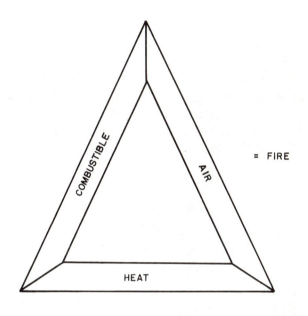

= FIRE

COMBUSTIBLE

AIR

HEAT

FIRE PREVENTION

Since Security Personnel are usually first on the scene when a fire breaks out and since man has never won mastery over fire, and since fire is so deadly, it needs closer study than most hazards. It is to the advantage of Security Personnel to be so familiar with fire, fire prevention, and fundamental fire fighting that actions become automatic when fire is discovered. Posters showing the kinds of fires and the classes of fires with approved types of extinguishers and how to operate them are available. Security Personnel should see to it that they are posted in as many places as possible. The use of the wrong type of fire extinguisher on a fire may cause it to burn even heavier.

THREE KINDS OF FIRES

1. The Live Combustion Fire. This fire has an open flame and generates a great deal of heat and causes smoke.
2. The Slow Combustion Fire. This fire is slow burning, has little flame, puts up little smoke, but generates a great deal of heat.
3. Spontaneous Combustion. This fire starts immediately and everything burns rapidly.

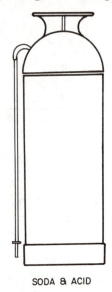

SODA & ACID

FOUR CLASSES OF FIRE

Class A — This is a dry fire. Paper — Wood — Textiles.
Class B — This is a fire from flammable Liquids or Grease.
Class C — This is a fire from Gas.
Class D — This is a fire from Chemicals or Metals such as Sodium and Magnesium.

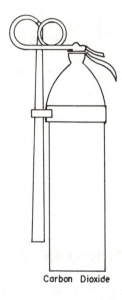

Carbon Dioxide

WHEN FIRE BREAKS OUT

The first thing to do when a fire is discovered is: **Report It!** Use the fire alarm box or the telephone to ask for help. The following should be included in the post orders:

— Security Personnel shall return to the fire scene after making sure the called for help can find the exact spot of the fire.

— Security Personnel shall give help and information to fire department personnel.

The supervisor of security, and management, shall be notified as soon as possible. Notes shall be made so that a good formal report can be made at a later date.

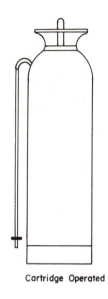

Cartridge Operated

SECURITY PERSONNEL MUST KNOW

1. All emergency telephone numbers.
2. The location of all fire alarms and fire extinguishers.
3. The types of fire extinguisher to be used and how to use it. Any fire department is happy to cooperate in teaching these basics.
4. The location of all fire doors, emergency exits, fire sprinkler valves, fire hydrants and fire hoses.
5. Specific and general duties as they apply to evacuation measure, protection of inventory and equipment, and fundamental fire fighting.

Security Personnel are not professional firemen. But, as the first line of defense, they must remain calm and collected. Others depend upon the actions taken and direction given by Security Personnel.

Prevention is the answer to protection from fire. Security Personnel will do well to make a check list of fire hazards that may be found when making rounds.

FIRE HAZARDS

1. Improperly identified fire extinguishers.
2. Fire extinguishers out of place or out of date or in poor condition.
3. Fire extinguishers blocked by furniture, equipment, or inventory.
4. Emergency exits blocked or hard to find due to poor lighting.
5. General hazards from heat.
6. Poor housekeeping practices such as piles of boxes, papers, etc.
7. Smoking in "NO SMOKING" areas.
8. Overloaded electrical systems.
9. Cars parked in fire lanes or too near fire hydrants.
10. Defective wiring and switches.
11. Open containers with flammable liquids in them.
12. Fire escapes blocked or otherwise obstructed.
13. Sprinkler system with no water in it.
14. Inventory or equipment stacked too close to sprinkler heads.

CHAPTER XVI

PUT IT IN WRITING

REPORTS

Security Personnel are the same as many other employees when it comes to making reports — they do not like to make them. They feel they are of little use, consume time, and seldom read. However, if they look at reports in a positive manner, they will realize their reports are the best defensive weapon they have.

Security Personnel reports are used by management in many ways. The main use is to verify the record of what happened during the shift. A clear and accurate report protects the employer as well as the individual. All reports should back up the Post Log, Post Journal, or Shift Report.

The Post Log, Post Journal or Shift Report is a history of the shift. It gives the next shift information to use. It gives dates, times, names, weather conditions and other data that may be vital to future legal proceedings. It will have information on the condition of weapons carried, lights, communication systems, the number of alarms answered, and the time and number of patrol rounds completed. It will also contain reference to any Incident Report that is being attached. It will also have instructions or recommendations for the shift that follows.

Any Security report, however given, should answer five questions, *Who, What, When, Where and How*. A sixth question may possibly be answered — *Why?*

Using the questions as they apply to the title of this chapter, ("Reports") will help Security Personnel make them a part of their conscious and sub-conscious minds.

Who makes reports? All Security Personnel must make reports. The outside patrol, the fixed post guard, the inside security desk, the special duty officer and anyone else working the security detail.

What is put in the report? Anything that has a bearing on, or what happens during the shift is placed in the Shift Report. This is the report from which other reports are made. Each report form to be filled in asks for specific information.

When is the report made? The Shift Report is an ongoing report. It is completed as events take place. All other reports are made as time permits. Reports made at the scene are the most accurate reports.

Where is the report made? At the duty post, if possible. If not, at the nearest place to physically make the report.

How is the report made? In person, if possible, on the radio or telephone if urgent. By telegraph, by express mail, or by handing to a supervisor. Many reports are given verbally and are then followed by a written report.

Why make a report? A report is made to keep the employer up to date on the security process. It is made so a permanent record of what happens on a shift is available for future use. It is made so action may be taken to continue, to correct, to change, or otherwise manage the security effort. The report, when properly made, protects the employer and also protects Security Personnel.

The oral or verbal report is used much of the time. It is direct, covers most of the details and is the easiest report to make. However, it is limited by the nature of the spoken word. It is a report made by one person to another person and often suffers changes when relayed to top management. The report can be made to grow or it can leave out details, when orally or verbally passed to others. For these reasons, it is always best to follow oral or verbal reports with a written report.

Written reports are made in many forms. The short note or memorandum is one method. The "fill in" form furnished by management is another method. The long narrative report is a third method. Most Security Personnel find themselves using each of these methods in making reports. The correct method or methods to use in making reports is the one asked for by your employer.

THREE THINGS TO REMEMBER
WHEN MAKING ANY REPORT

1. Be liberal with words in the report — use many words. It is much better to use too many words or paragraphs than it is to leave out the slightest detail. Be a tattle tale!
2. Write what you know. It is important to get it down on paper. Do not worry about spelling or style. The *Who, What, When, Where, and How* can be answered — sometimes the *Why* cannot be answered.
3. When ready to hand in the report, other than the Shift Report, make a copy. Make carbon copies or photo-

copy reports so they may be referred to at a later date.

Reports should always contain answers to the five or six questions *Who What, When, Where, How* and possibly *Why.*

What happened? The report should describe any incident from beginning to end. The facts, as seen, or found, are important to employee and employer. The incident report may need to be reviewed at a later time when testimony is being taken for insurance, court suit, or other purposes.

When did it happen? The time of day, the day, the month and the year must be included. When did it happen? When did security first become aware? When did security first report? When did help arrive? When did security return to post? When was it written down?

Where did it happen? The exact location where the incident took place is required. The room number, floor level, hallway, kitchen, lobby, or other location must be pinpointed if the incident happened inside a building. If outside, and easily identified, the location may be noted as East parking lot, front entrance, side door, swimming pool or loading dock. If the incident happened away from an easily identified place, it is a good idea to make a diagram. Using permanent objects for markers and making a few measurements to the place of the incident a sketch can be drawn for reference. For example, a thief is caught going over the property fence. A measurement from a permanent corner to the place where the capture was made pinpoints the spot.

Who was there? Every person connected with the incident needs to be identified. Names and addresses of witnesses are valuable. Security Personnel, police, fire and medical personnel, victims, investigators and others who were on the scene should be named in the report.

How did it happen? Any information available to help find the cause of the incident must be noted. "The customer stepped on a moving escalator and slipped down." "The padlock on the rear gate was cut off and entry made at that point." "It was raining and the truck slid off the entrance road into the ditch." "The cashier was ringing up a no sale and taking bills from the register." Survey the scene, and make mental notes to be written down later. The *How* it happened portion of the report takes careful thought and common sense.

Why did it happen? This question is hard to answer. Why the lightning struck or the hurricane blew cannot be answered. The reasons *Why* something happens is usually an opinion. If in the report, it should be noted it is an opinion.

Security Personnel who make good oral and written reports receive more credit than they know from their supervisors. It is one of the best ways to make a silent impression on an employer.

Exhibit 1 has samples of report forms. Permission to copy will be given if a written request is sent to the author.

WHEN MAKING YOUR REPORT —
INCLUDE THE FOLLOWING :

WHO —

Who was involved?

Names — Ages — Addresses — Telephone Numbers

Who were witnesses? — What were their jobs?

WHAT —

What happened?

What took place? — Was it an accident? — Was it intentional?

What caused it? — Write down the events from beginning to end as you recall them.

— Use as much space as you need. — This is the time to be a Tattle Tale.

WHEN —

When did you first know? — Time of day? — Day of week? — Month? — Year? — When were you notified? When did you arrive on scene? — When did you report the the event? — When did help arrive? — When did you leave the scene?

WHERE? —

Where did the event occur? — Where were you when it happened? — Where were the persons involved? Where did help come from? — Where did you go after the event was reported?

HOW —

How did it happen? How did you learn about it? How did you handle the situation? — How many others were involved? — How was it caused?

WHY —

The motivation, if any, behind the incident. Why do you think it happened?

THE ABOVE ARE WORDS THAT MAY HELP YOU IN MAKING YOUR REPORT. USE THEM, OR OTHERS, IN MAKING A REPORT THAT WILL EXPLAIN TO OTHERS WHAT YOU DID AND WHAT OTHERS DID REGARDING THE ENTIRE INCIDENT. BE A "NEWS" REPORTER!

DAILY REPORT FORM

Check each item. If YES is checked, write your finding in comment column. Write on back of sheet if necessary.

WRITE OR PRINT SO IT CAN BE READ

DATE _____ LOCATION _____

SHIFT _____ POST _____

ON YOUR SHIFT DID YOU	YES	NO	COMMENTS
Have An accident?			
Have Visitors?			
Have Intruders?			
Have A Fire?			
Check Safe?			
Find Open Window/Door?			
Find Damaged Equipment?			
Check Fire Extinguishers?			
Find Vandalism Signs?			
Have Lighting Problems?			
Check Electrical Boxes?			
Check Fuel Supplies?			
Check Perimeters?			
Find Safety Violation?			
Find Physical Damage?			
Make All Rounds?			
File An Incident Report?			
Find Open Locks?			
Find Theft Signs?			
Have Other Problems?			

_____
Signature

_____
Badge Number

SHIFT REPORT

This is a runing report. Always use the next number in order in the "Report Entry Column". Be sure to complete the "TIME", "REPORTED BY", and "ACTIVITY" column.

WRITE OR PRINT SO IT CAN BE READ

DATE _____ LOCATION _____

SHIFT _____ POST _____

HOURS

FROM: _____A.M./P.M. TO: _____ A.M./P.M.

REPORT ENTRY
    NUMBER   TIME — AM/PM REPORTED BY   ACTIVITY

IF A SEPARATE REPORT IS BEING MADE OF ANY ACTIVITY USE THE REPORT ENTRY FROM ABOVE TO IDENTIFY IT.

REPORT OF PERSONAL INTERVIEW

This is a report of an interview with a person held while on duty. Be sure you get all information that is indicated on the form and any you wish to add.

WRITE OR PRINT SO IT CAN BE READ

DATE _____ LOCATION _____

SHIFT _____ POST _____

Where did you hold interview? _____

_____

What reason did you have to hold the interview? _____

_____

Name or names of persons interviewed with addresses and telephone numbers: _____

_____

_____

If employee, what identification? (Get department and I.D. Number and reason for being in the area) _____

_____

_____

If car used, get: Make _____ Model _____ Color _____

Brief Description of each person — include: Race _____

Age _____ Sex _____ Height _____

Weight _____ Hair _____

Any outstanding characteristic _____

_____
Signature

_____
Badge Number

PATROL REPORT

This is to report any hazard or safety violation you find during patrol. Make a detailed report.

WRITE OR PRINT SO IT CAN BE READ

DATE _____ LOCATION _____

SHIFT _____ POST _____

KIND OF HAZARD OR VIOLATION

Number — (Use in giving location and problem)

1. Aisles, Stair, Walkways, Floors
2. Doors, Windows, Other Entry or Exit Points
3. Lights and/or Lighting
4. Safety Equipment
5. Vehicle Parking and/or Operations
6. Electrical
7. Plumbing
8. Fuel Supply
9. Fire Hydrants and/or Extinguishers
10. Fences and/or Gates or Walls
11. General Outside Area
12. Other Hazards or Violations

NUMBER _____ FROM ABOVE IS LOCATED AT _____

_____

AND THE PROBLEM IS _____

_____

OTHERS INVOLVED WHO HAVE ___ OR SHOULD BE ___

NOTIFIED, (CHECK ONE) INCLUDE: _____

_____

ACTION TAKEN _____

RECOMMENDATION _____

_____
Signature

_____
Badge Number

BURGLARY REPORT
(Complete Each Blank)
WRITE OR PRINT SO IT CAN BE READ

DATE _____ LOCATION _____

SHIFT _____ POST _____

(Use back of sheet or use another sheet if needed)

PLACE WHERE BURGLARS ENTERED _____

_____

PLACE WHERE BURGLARS DEPARTED _____

_____

ITEMS STOLEN _____

_____

ESTIMATED VALUE STOLEN ITEMS_____

BURGLARY FIRST REPORTED _____ A.M/P.M

BURGLARY REPORTED BY _____

POLICE/SHERIFF (Name) NOTIFIED _____ A.M./P.M.

POLICE/SHERIFF (Name) _____

RESPONDED TO SCENE _____ A.M./P.M.

LAW INVESTIGATOR (Name) _____

DESCRIBE BURGLARY AND BURGLARY SCENE _____

_____

_____

_____

Signature

_____

Badge Number

ACCIDENT REPORT

(Complete Each Blank)

WRITE OR PRINT SO IT CAN BE READ

DATE _____ LOCATION _____

SHIFT _____ POST _____

(USE BACK OF SHEET OR USE ANOTHER SHEET IF NEEDED)

Type of Accident **(Circle One) Person Vehicle Property**

Location Of Accident _____

_____

Accident First Reported By _____

At _____ A.M./P.M. The Accident Happened

At _____ A.M./P.M. Actual ____ Estimated ____

Accident Reported To Following At Time Indicated:

Medical Facility (name) _____ A.M./P.M.

Police/Sheriff (name) _____ A.M./P.M.

Your Superior (name) _____ A.M./P.M.

Property Owner (name) _____ A.M./P.M.

Fire Department (name) _____ A.M./P.M.

Other (name) _____ A.M./P.M.

Names of Victims with addresses and telephone numbers if possible — be sure to identify location employees. _____

_____

Describe Accident Situation From First Report Until You Left The Scene (Use attached WHO, WHAT, WHERE, WHEN, Etc.)

_____

_____

_____

_____

Signature

_____

Badge Number

FIRE INCIDENT REPORT

(Complete Each Blank)

WRITE OR PRINT SO IT CAN BE READ

DATE _____ LOCATION _____

SHIFT _____ POST _____

(Use back of sheet or use another sheet if needed)

Exact Location of Where Fire Started: _____

_____

Fire First Reported By _____

Time of First Report of Fire _____ A.M./P.M.

Time Fire Department Notified _____ A.M./P.M

Time Fire Department Arrived _____ A.M./P.M.

Time Police/Sheriff (Name) Notified _____ A.M./P.M.

Estimate of Fire Damage _____

Describe fire from first report until you left the scene (Use the WHO, WHAT, WHERE, WHEN, HOW AND WHY SHEET)

_____

_____

_____

_____

_____

_____

_____

Signature

_____

Badge Number

THEFT INCIDENT REPORT

(Complete Each Blank)

WRITE OR PRINT SO IT CAN BE READ

DATE _____ LOCATION _____

SHIFT _____ POST _____

(Use back of sheet or use another sheet if needed)

Describe the stolen property _____
_____

Theft was from what Person and/or Department _____
_____

Theft reported by _____

Theft reported to _____

Time theft reported _____ A.M./P.M.

Police/Sheriff (Name) notified _____ A.M./P.M.

Law Investigator (Name) _____

Estimated value stolen property _____

Suspected type of theft _____
               Burglar   Shoplifter   Employee

Describe the theft and the theft scene _____
_____
_____

(Use the WHO, WHAT, WHERE, WHEN, WHY AND HOW SHEET)

_____
Signature

_____
Badge Number

SECURITY GUARD GLOSSARY

ACCIDENT — An unexpected and usually undesirable happening or mishap.

ARREST — Detention of a person against their will.

BOMB — An explosive weapon that can be detonated by predetermined means.

BURGLAR — A person who enters a place with intent to steal.

CONDUCT — The way a person acts or behaves.

CRIME — An unlawful act or acts.

DEMEANOR — The manner in which a person handles him or herself.

DETENTION — The act of detaining a person.

EMPLOYEE — A person who works for another for compensation.

EMPLOYER — A person or concern that employs people for compensation.

FIRE — A burning substance that gives off heat and light.

FIREARM — A weapon using explosive material to propel a missile.

GUARD — One who guards and is a protector.

GUARDING — The act of one who prevents and protects.

HAZARDS —

   (Human) Perils to life and/or property caused by people.

   (Natural) Events in nature that place persons and/or property in peril.

I.D. — Common abbreviation for identification.

IMAGE — The way a person appears in the minds of others.

JURY — A group of persons in a body sworn to judge and give a verdict on a matter.

JUVENILE — A young person not yet an adult.

KEY — An instrument used to open a lock.

KLEPTOMANIAC — A person who has an obsession to steal without apparent reason.

LEGAL — Within the authority of the law.

LOG — A book in which records are kept.

MOB — A disorderly crowd.

MOTIVATION — The incentive for acting.

NATIONAL — Belonging to a Nation as a whole.

NATURAL — Present in, or produced by nature.

OFFICER — A person who holds an office.

OFFICIAL — Authorized by proper authority.

PATROL — The movement about an area for the purpose of observing and/or for security purposes.

PERIMETER — The outside boundary of a security patrol.

PILFERAGE — Stealing items from an area by one inside the area.

PREVENTION — Steps taken to keep an event from happening.

PROTECTION — The act of guarding life and/or property.

PROSECUTION — Taking action against another.

REPORT — An account of events during a period of time.

ROBBERY — The unlawful act of taking property.

SABOTAGE — Underhanded act to stop normal operations.

SHOPLIFT — Stealing goods from a store display.

STEAL — Taking a valuable without the right or permission to do so.

TRAINING — Action taken to prepare a person for a special task.

TRESPASS — To enter another's property without permission.

THEFT — The act of stealing.

UNIFORMED — Dressed in a distinctive manner.

VANDALISM — Willful destruction of private or public property.

V.I.P.'s — (An informal abbreviation for Very Important Person) The Security Guard is a Very Important Person. The Security Profession is a Very Important Profession.

WITNESS — One who is called upon to testify in court.

INDEX

ORDER FORM

Security Seminars Press
1204 S.E. 28th Avenue
P. O. Box 70162
Ocala, FL 32670

Yes, please send me _____additional copies of
SECURITY GUARD by David Y. Coverston at $22.00
postpaid. Florida residents add 5% sales tax.

Name _____

Address _____

City _____ State _____ Zip _____

☐ I am also interested in other training materials
relating to security operations.

☐ Send information about the book "Security Training and Education".

☐ Send information about the book "Security for
Senior Citizens".

☐ Put me on your mailing list.

☐ I am interested in a security newsletter.

ABOUT THE AUTHOR

David Y. Coverston, author of *SECURITY GUARD*
has had a long and distinguished career in a variety of
professions. President of Florida Security & Investigators Association, he has memberships in the American
Society for Industrial Security, the Professional Security Officers International and the Florida Peace Officers Association. He is presently serving on the subcom-
mittee for security officer education, Private Security
Advisory Council for the Florida Department of State.
Past experiences in the field includes responsibilities
for security at a private resort in Sarasota, Florida,
Security Chief at Manatee Community College in
Bradenton, Florida, and Director of Security for World
Golfers Association Tournaments. A native of Okla-
homa, he currently resides in Ocala, Florida.